COMMUNICATING IN THE AGRICULTURAL INDUSTRY

Delmar is proud
to support FFA activities

Join us on the web at
agriculture.delmar.com

COMMUNICATING IN THE AGRICULTURAL INDUSTRY

RUSSELL A. GRAVES

THOMSON

DELMAR LEARNING

Australia Canada Mexico Singapore Spain United Kingdom United States

THOMSON
DELMAR LEARNING

Communicating in the Agricultural Industry
Russell A. Graves

Vice President, Career Education Strategic Business Unit:
Dawn Gerrain

Director of Editorial:
Sherry Gomoll

Developmental Editor:
Andrea Edwards

Editorial Assistant:
Rebecca Switts

Director of Production:
Wendy A. Troeger

Production Manager:
Carolyn Miller

Production Editor:
Kathryn B. Kucharek

Director of Marketing:
Donna J. Lewis

Channel Manager:
Nigar Hale

Cover Images:
Getty One

Cover Design:
Dutton & Sherman Design

For permission to use material from this text or product, contact us by
Tel   (800) 730-2214
Fax   (800) 730-2215
www.thomsonrights.com

Library of Congress Cataloging-in-Publication Data:

Graves, Russell A., 1969–
   Communicating in the agricultural industry / Russell A. Graves.
      p. cm.
   Includes bibliographical references and index.
   ISBN 1-4018-0889-1
   1. Communication in agriculture—Study and teaching (Secondary)—United States.   2. Agricultural communicators—Education—United States.   I. Title.
S494.5.C6G73 2005
630'.1'4—dc22
                                                      2003055548

NOTICE TO THE READER

Publisher does not warrant or guarantee any of the products described herein or perform any independent analysis in connection with any of the product information contained herein. Publisher does not assume, and expressly disclaims, any obligation to obtain and include information other than that provided to it by the manufacturer.

The reader is expressly warned to consider and adopt all safety precautions that might be indicated by the activities described herein and to avoid all potential hazards. By following the instructions contained herein, the reader willingly assumes all risks in connection with such instructions.

The Publisher makes no representations or warranties of any kind, including but not limited to, the warranties of fitness for particular purpose or merchantability, nor are any such representations implied with respect to the material set forth herein, and the publisher takes no responsibility with respect to such material. The Publisher shall not be liable for any special, consequential, or exemplary damages resulting, in whole or part, from the reader's use of, or reliance upon, this material.

To Glenda, Dee, Bubba, and Larry. No one could ask for better sisters and brothers. You have helped make me who I am today.

CONTENTS

PREFACE

Communicating about agriculture is not a concept shrouded in mystery. Instead, it is commonsense knowledge that you have learned throughout your schooling in language, history, computer, and math classes. This book challenges you to take those lessons learned and apply them to the science and art of communicating.

This book teaches you about communicating the good message about one of the most important industries on earth today: the agricultural industry.

It is easy to see why having strong skills in communicating about agriculture is so important. People who have an intimate knowledge of agriculture and can communicate effectively are comparatively rare.

Increasingly, we are living in an agriculturally ignorant society. As the nation becomes more urbanized, people are losing touch with the land that feeds them. As a result, food safety scares are becoming increasingly common. People are concerned about the quality and safety of their food, and rightfully so. Although they cannot boycott food altogether, they can rally against certain sectors of the food-producing industry such as the beef industry. When boycotts of a sector of the agricultural industry occur, everyone in the industry suffers because of the close ties that one sector has to another. Everyone suffers, because consumer confidence is shaken—consumers who ultimately drive the price and consumers whom the agricultural industry must strive to please. As consumer confidence is shaken, a gap occurs between agriculture and the rest of society because people get the impression that some agricultural practices are bad for their families and bad for the environment.

This book is meant to help bridge that gap. If agriculturists can communicate effectively about their industry, then everyone wins, and consumer confidence remains high. The book, however, is not meant to be the authority on communicating in the agricultural industry. Its purpose is to give you a reference point for further investigation into the discipline. The sections on writing should be supplemented with other books on writing, along with a good dose of practice. The same concept applies to the sections on speaking, photography, and graphic design. This book is your starting point to expand your knowledge of all of these disciplines.

FEATURES OF COMMUNICATING IN THE AGRICULTURAL INDUSTRY

Each chapter is laid out in a logical fashion, moving from the theory and concepts of the topic to practical usage of that topic. From one chapter to the next, skills are introduced that build on the lessons learned earlier.

Each chapter ends with study questions to help you fully grasp the lessons and concepts each chapter imparts. Enhancement activities allow you to learn and perfect practical skills that pertain to agricultural communications, including news release writing, feature story writing, photography, Web page design and construction, public speaking skills, and layout. At the end of each chapter, you will find Web site listings and bibliographical information to help you further investigate the topics each chapter offers.

The appendix features a comprehensive list of jobs relating to agricultural communications and a list of universities where you can pursue a course of study in agricultural communications.

At the end of the textbook, you will find a glossary of terms used in each chapter to help you fully understand the concepts presented.

- **Chapter 1.** This chapter discusses consumer trends and the need for effective agricultural communications. It puts the discipline of agricultural communications in context with a brief history of the industry, complete with important dates. Visual and writing trends are discussed to put into context why magazines and other publications have a certain "look" and change that look periodically to keep up with the changing consumer markets.
- **Chapter 2.** This chapter discusses ways to build strong credibility by conducting solid research. The lesson on research discusses how to take a broad topic and narrow it to a usable thesis. Techniques are discussed that you can employ to get information you may need for a story. You will learn the importance of double-checking facts and properly attributing information. Finally, plagiarism is discussed.
- **Chapter 3.** This chapter moves from theory to practical application. You will learn how to apply what you have learned in English and language classes to real-world situations. You will learn some of the pitfalls associated with ineffective communication and, more important, how to avoid them. You will also learn how to prepare news releases and feature stories and understand what determinants make good news stories.
- **Chapter 4.** In this chapter, you will learn how to recognize the importance of effective electronic communications. A discussion of how the Internet and e-mail works is appropriate for this chapter, as a basic knowledge of how the World Wide Web works is essential so you can plan and implement Web sites effectively. You will also learn design concepts that pertain to Web site construction and how to market the Web site once it is up and running.
- **Chapter 5.** One of the elemental disciplines of agricultural communications is the ability to communicate with pictures. This chapter teaches how to take better pictures using simple techniques. It also discusses the difference between a digital and a film camera so that you can make a decision as to which

type of camera is right for you. Finally, it provides some techniques on how to illustrate a story.

- **Chapter 6.** In this chapter, you will explore ways to help you prepare for the age-old challenge of speaking in public and making the most of your time in the public eye. You will see how to plan and implement extemporaneous speeches as well as prepared public speeches. The chapter features a prepared speech that was written by a student for competition. As you progress through the chapter, you will also get hints on how to use your time wisely, learn about the role of spokesperson, and discover how to add visuals to presentations.
- **Chapter 7.** Good layouts and designs do not happen haphazardly. They are thought out and follow a prescribed set of time-tested rules that outline how elements of a layout, such as text and graphics, should be arranged. You will learn about basic layout considerations such as alignment, contrast, balance, and proximity. Using a computer for saving time is discussed. The introduction to the electronic darkroom will help you make the most of the artwork you have to work with.
- **Chapter 8.** In this final chapter, you will learn about the entrepreneurial opportunities that await you as a freelance agricultural communicator. The chapter discusses the existing market for writers and photographers interested in communicating about the agricultural industry and what kind of compensation exists for those willing to try the business. You will also learn how to make contact with editors and what you need to do to build and maintain strong relationships with people in charge of making editorial decisions. Finally, you will learn how to write a query letter that gets the attention of people buying creative content.

This book is about learning how to communicate in the agricultural industry. As such, each student using this book is charged with taking lessons he or she has learned in computer science and language arts and putting them to use. By honing your skills in these disciplines, you serve a valuable service to agriculture through training to pass on a positive message about one of our nation's most crucial industries.

This book is not meant to serve as the definitive word on the subject of agricultural communications. It is meant as a starting point to give you direction in contributing creative and accurate content to the national dialogue about agriculture. That is why, at the end of each chapter, Web links are provided that will point you to sites that will enhance your knowledge of each chapter and lead you into further explorations about the topics presented in this book.

By using all of the features of this book, you will develop a firm grasp of the concepts of agricultural communications. Through repeated practice and careful guidance, you will master these concepts.

As agriculturists, we are stewards of the land and caretakers of animals that were domesticated hundreds of years ago. As communicators, we are caretakers of an industry—one of the most important industries in the history of humankind. It is our duty as students and teachers of agriculture to learn, as efficiently and thoroughly as we can, the skills needed to tell a positive story about agriculture through words, pictures, and Web site content. This book will help you in that goal.

ABOUT THE AUTHOR

Russell Graves is an agricultural science instructor at Childress High School in Childress, Texas and award-winning photographer and writer whose credits include many magazine covers and books on wildlife and agricultural subjects. He has been a keynote speaker for teachers' groups, chambers of commerce, outdoor communicators, and youth gatherings. Raised in Dodd City, Texas, Graves brings small town humor and values to his presentations. A life spent in close contact with the natural world makes his speeches and photography unique.

Graves is deeply committed to educating people about the value of wildlife and its conservation and agriculture. He was nominated in 1999 and 2000 for the coveted Conservation Educator of the Year by the Texas Section of the Wildlife Society, and the Texas Association of Soil and Water Conservation Districts. In the spring of 2000, he was named Region 1 Conservation Educator of the Year by the Texas Association of Soil and Water Conservation Districts and won the Area 1 FFA Association's Agriscience Teacher of the Year title. In the summer of 2001, he was named Texas Agriscience Teacher of the Year and was a finalist for the national honor. Projects under his tutelage have gained national prominence, such as the Black-Tailed Prairie Dog Research Project, in which his students conducted research on plant diversity on burrowing habits of prairie dogs. The project, which gained much regional press attention, was also featured by the Microsoft Corporation in an educational newsletter and on their Web site.

He is also a National Finalist for Agriscience Teacher of the Year in 2003 and was recently named a national winner of the H.O. Sargent Award by the National FFA Organization. The H.O. Sargent Award is given annually to people who promote cultural diversity in the FFA.

Graves's work has appeared in the pages and on the covers of numerous state and national publications and has been featured in books on agricultural and wildlife subjects. He maintains a constantly expanding file of wildlife, nature, outdoor, and agricultural images. In addition to marketing his images, he is represented by AGStockUSA, the world leader in providing the highest-quality and greatest variety of agricultural, produce, and livestock photography for advertising, corporate communications, and editorial usage.

Graves's students have won seventeen state championships and four national championships since 1998. His students have won speaking contests, agriscience fair contests, career development events, and a variety of other award areas sanctioned by the FFA.

ACKNOWLEDGMENTS

I thank Dane Fuller for all of his input in the development of this manuscript. In addition, thanks to my students at Childress High School as well as my wife, Kristy, and daughter, Bailee, for their patience through the development of this book.

Delmar and the author wish to express their thanks to the following individuals who devoted their time and professional expertise to reviewing this text manuscript.

Amanda J. Brown
    Station Camp High School
    Gallatin, Tennessee

Rachel A. Zabel
    New Richmond High School
    New Richmond, Virginia

Jeff McGinnis
    Ropes High School
    Levelland, Texas

Craig Lister
    Marysville High School
    Marysville, Kansas

Mark Lalum
    Flathead High School
    Kalispell, Montana

Internet Disclaimer

The author and Delmar affirm that the Web site URLs referenced herein were accurate at the time of printing. However, due to the fluid nature of the Internet, we cannot guarantee their accuracy for the life of the edition.

THE AGRICULTURAL COMMUNICATIONS INDUSTRY: AN OVERVIEW

OBJECTIVES

After completing this chapter, you should be able to:

- identify major changes in the agricultural industry
- understand the need for competent agricultural communicators
- identify historic events in agricultural communications
- understand the future of agricultural communications

KEY TERMS

agricultural literacy   trends

OVERVIEW

Chances are that when most people think about the agricultural industry, they envision neatly cultivated row crops or green pastures full of fat cattle (Figure 1-1). Traditionally, serene, pastoral scenes are the rule for agriculture in America, and that still applies today. But over the past half-century, agriculture has seen major changes. No longer is it limited to production (Figure 1-2). In fact, the field of agriculture encompasses more disciplines now than it ever has before in its long and storied history as America's oldest industry. Researchers, geneticists, entomologists, botanists, and other highly specialized positions now are part of the agricultural industry.

Communication specialists have an important place in agriculture too. In this chapter, you will learn how communicators are important to

FIGURE 1-1   Traditional farm scenes are still common, but agriculture has seen major changes over the past half-century. (*Courtesy USDA/ARS*)

agriculture and what services they provide to a dynamic industry. In addition, you will learn how agricultural communications isn't a new field. It is a discipline that has evolved into a powerful tool that disseminates information on the world's most crucial industry. Moreover, you will learn about **trends** in the agricultural communication industry and how those trends are likely to affect agriculture. Finally, you will consider the future of agricultural communications and how the field will become one of the most important sectors of the agricultural industry.

FIGURE 1-2   Today agriculture encompasses more disciplines than just production. (*Courtesy USDA/ARS*)

THE AGRICULTURAL COMMUNICATIONS INDUSTRY: WHY THE NEED EXISTS

In a 1988 study by the National Academy for Sciences, the researchers discovered that people's general understanding of agriculture in the United States (Figure 1-3) is low. Studies in Kansas and Arizona back up the academy's findings by showing that most children, teenagers, and even adults have little knowledge of basic agricultural concepts. In other words, the public generally has a low level of **agriculture literacy**—that is, an understanding of the processes by which food and fiber are brought from the farm to the consumer. Is it important for people to have an understanding of agriculture? Yes!

In the Arizona study, researchers found that adults have a generally negative view of agriculture and believe that the industry has a detrimental effect on wildlife, food safety, and water (Figure 1-4). Compounding the problem of the public's opinion on agricultural issues is the constant barrage of negative agriculture-related stories by the mass media.

If you follow the news closely, you'll see that stories appear regularly about food safety, pesticide concerns, and the effects of cattle grazing on

FIGURE 1-3   A National Academy of Sciences study shows that Americans have little understanding of agriculture. (*Courtesy Russell A. Graves*)

rangelands. Rarely are these news stories favorable to agriculture (Figure 1-5). Instead, they often are reactive pieces that point out the negative consequences that occasionally occur in the agricultural industry. And when these stories reach the newsstands or radio or TV, people may become alarmed. This brings about an important point. Unlike some other industries, agriculture is one that we cannot live without. Our whole food and fiber system, as well as a big portion of our gross state and national products, relies on the fruits of agricultural labor. Not only does our health depend on agriculture, but our national security relies on it too. Imagine for a moment how vulnerable our country would be if we had to rely on another country for our food and fiber products. Do you think we'd still be a world superpower? It is a scary thought isn't it?

This background brings us around to the central point of this textbook: the strong need for competent and agricultural communicators (Figure 1-6). All over the country, newspapers, magazines, radio, and television stations regularly run agricultural features, but many do not have anyone on staff versed in the agricultural sciences. They are proficient as communicators, but it can be successfully argued that they need expertise in agriculture to report effectively on the nation's most vital industry. What do you think?

FIGURE 1-4   Studies show that American adults have a generally negative view of the effects of agriculture on resources such as water. (*Courtesy Russell A. Graves*)

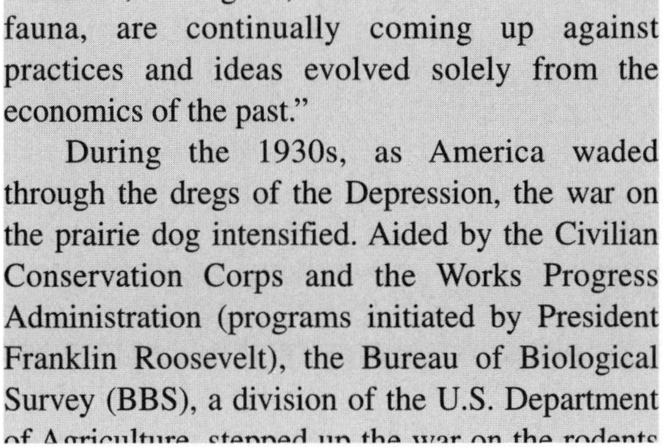

scientific, ecological, and aesthetic — of wild fauna, are continually coming up against practices and ideas evolved solely from the economics of the past."

During the 1930s, as America waded through the dregs of the Depression, the war on the prairie dog intensified. Aided by the Civilian Conservation Corps and the Works Progress Administration (programs initiated by President Franklin Roosevelt), the Bureau of Biological Survey (BBS), a division of the U.S. Department of Agriculture, stepped up the war on the rodents

FIGURE 1-5   Newspapers are full of stories unfavorable to agriculture.

FIGURE 1-6   Today there is a strong need for people who can communicate positively about agriculture. (*Courtesy USDA/ARS*)

The agricultural industry benefits from a corps of progressive, proactive, and competent agricultural communicators because they know how to report the good news of the industry in an unbiased and factually accurate manner. This is the purpose of this textbook and the lessons you will be studying.

All over the country, the opportunity to speak and write positively about agriculture exists. Since you are studying agricultural communications, chances are that you have had or will have other classes in traditional agricultural subjects, such as animal and plant production, genetics, or natural resource topics. Study of agricultural communications will build on what you have learned in computer science, English, and other related classes to help you make a difference in agriculture.

Out of the need to communicative effectively and positively about agriculture, agricultural communication programs in high school and college were born. By being a part of agricultural communications, you are helping to advance an industry that has helped make the United States great.

THE HISTORY OF THE AGRICULTURAL COMMUNICATIONS INDUSTRY

The history of agricultural communications helps to put current trends and practices in perspective and illustrates a legacy of accomplishment.

Like many other industries, the development of agricultural communications in the United States goes back to the colonial days when the society was predominantly agrarian. There are definable milestones that trace the changes in agricultural communications since the early days of the United States (Figure 1-7). The abbreviated list that follows was reported and adapted by Texas Tech University's Agricultural Communications Department.

1810—The first agriculture magazine, *Agricultural Museum*, was established in the District of Columbia.

1819—The magazine *American Farmer* was released in wide circulation.

1837—Reporter Horace Greeley wrote, "Go West, Young Man," and encouraged people to travel and settle the western territories.

1842—*American Agriculturist*, the nation's oldest farm magazine with the same name, was established in New York.

1862—The Morrill Act was passed by the U.S. Congress, establishing the nation's land grant university system (Figure 1-8).

1873—*Drover's Journal* was founded in Chicago (Figure 1-9).

1877—*Farm Journal* was established in Philadelphia.

1885—Over 172 agriculture journals were published during this year.

1894—The first *Yearbook of Agriculture* was issued by the U.S. Department of Agriculture.

1902—*Successful Farming* magazine was established in Des Moines, Iowa.

1905—The first class in agricultural journalism was offered at Iowa State University.

1908—The University of Wisconsin, Madison, created an agricultural journalism department.

1915—The first weather and crop reports were broadcast in Morse code over the radio at the University of Wisconsin.

1920—Iowa State University offered the bachelor of science degree in agricultural journalism.

1921—The American Agriculture Editors Association was organized.

1921—The first vocal weather and crop reports were broadcast over the radio from the University of Wisconsin.

1928—Seven U.S. universities offered courses in agricultural journalism.

1928—This year saw over 600 agriculture magazines in publication.

1944—The National Association of Farm Broadcasters was organized.

1952—The Newspaper Farm Editors of America was established.

1955—The circulation of farm magazines reached 29 million.

1964—The magazine *Beef* was established in St. Paul, Minnesota.

1970—Agriculture Communicators of Tomorrow, an organization for college students interested in agricultural communications, was established.

(continued)

1974—The Livestock Publications Council was established.

1990—In this year, there were over 102 agriculture newspapers, 440 magazines, 1,025 AM radio stations, 803 FM radio stations, 14 state and 3 regional radio networks, and 5 television stations specializing in agriculture.

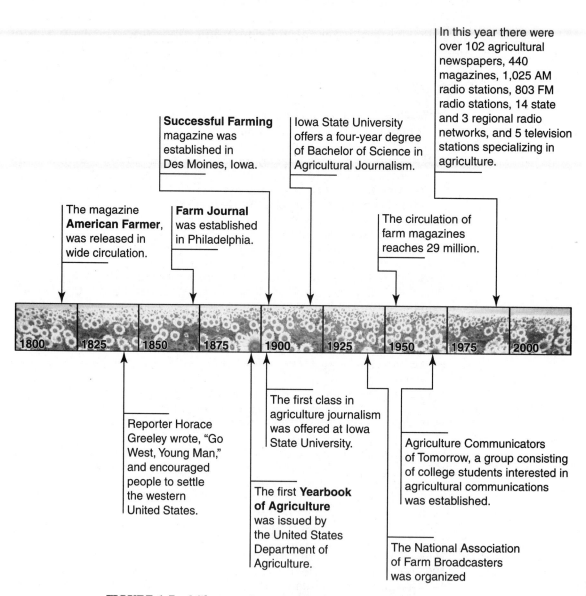

FIGURE 1-7   Milestones in agricultural communications.

FIGURE 1-8   The Morrill Act, establishing America's land grant university system, was passed in 1862. (*Courtesy USDA*)

FIGURE 1-9   Since the first agricultural magazine was introduced, the industry has spawned a number of special interest publications. (*Courtesy Dally Times Inc.*)

INDUSTRY TRENDS

A trend, in the consumer sense, is simply a pattern that people follow. You do not have to look very hard to discern trends in music, clothing, movies, and television shows. Trendy people and things drive our popular culture.

Why? Trendiness sells. Retailers and others in business know that people are always looking for something new and fresh. Look at computers as an example. For years, people had only one choice when it came to the color of their computer: white (or a variation of it). Then Apple Corporation started a trend by making its Macintosh line of computers in different colors. As sales of Macs jumped, other computer manufacturers joined the emerging trend, and now you can buy a computer in just about any popular color.

Our capitalistic society demands that products and their appearance constantly reinvent themselves. Industry competition forces companies to be a little bit different from their competitors every time they release a new product in order to stand out from the rest and be noticed. Change also helps to ensure survivability in the marketplace.

Trends have their own life cycle. For example, bell-bottomed pants were popular for a few years in the late 1960s and 1970s. After almost two decades in exile, bell-bottoms are in style once again.

Agricultural communications is no different. Just like consumer products, the products of agricultural communications evolve. Written and visual content changes yearly—sometimes subtly, sometimes not.

WRITING TRENDS

The art of writing is a continually evolving process. At one time, humans could not write at all. Now we use words to inform, persuade, and sell. Literacy is one of the most important tools that we have in our intellectual arsenal.

Not only has writing evolved in substance; the style of writing has also changed drastically over time. William Shakespeare's writing is a vivid example of how the style of writing and use of language has changed over the past few hundred years. It has even changed drastically over the past 100 years. The writing of pioneers during the last half of the nineteenth century was much more colorful, with more vivid imagery than today's average letter is. Even over the past two decades, writing has changed.

People now want information in small, easy-to-digest bits of information. Writing now, especially in the popular press and for the Internet, is pithier and faster paced than perhaps ever before. News outlets are challenged to provide relevant information in a fast-paced format. They know that if it takes a reader too long to get to the point of the article, that person is likely to choose a different paper or another Web site to find information.

Magazines are also following the same trends. In the past, feature articles were a flowing commentary that advanced from the beginning to the end of the piece. Although some magazines still follow that format, many publications now break articles into subarticles of 300 to 400 words that all fit together to form a single piece.

Of course, some media outlets still use traditional writing styles, and some with great success. Nevertheless, the trend for shorter pieces is evident.

Why has this trend toward more concise copy come about? Although the factors are many, the underlying truth is that writing is tighter now because of the competition.

For years, the only source of news was word of mouth. Then newspapers became popular. Next, radio was introduced. It then had to deal with the introduction of television, which evolved into multichannel information devices (Figure 1-10). Somewhere in the mix, magazines came of age.

FIGURE 1-10   Television has long been a means of disseminating information like commodity reports to agricultural producers—such as sunflower producers. (*Courtesy USDA/ARS*)

And now e-mail and the Internet have become major players in disseminating news.

There are now countless ways to get news and information. News bureaus and advertisers recognize that fact and have evolved to meet the demand for quick news with just the facts and a source for analysis for those who want to know more. In the information game, time is money. In other words, your time guarantees them money.

Moreover, space is money. For every square inch of newspaper space with news, there is another square inch of advertising space meant to generate income for the publication. Papers must put as much information in an edition as they can without sacrificing ad space. Therefore, relatively short and factually accurate articles are the rule. The same principle applies to magazines, television shows, radio programs, and Web sites.

VISUAL TRENDS

Like writing, visual trends in publications, namely photography, have also changed and continue to evolve. Photography used to be a straightforward process. For the most part, what you saw is what you got. As a result, most photographs from thirty years or so ago and beyond had a big depth of field (the amount of space in front of and behind a subject that is in focus) and a two-dimensional perspective; they were accurate representations of what the camera saw (Figure 1-11). For journalistic types of photography, where capturing a slice of time was important, this type of image making was fine.

Like all other aspects of our capitalistic culture, competition has caused companies to be bold in their advertising when it came to their advertising and their corporate identities. As a result, cameras and the techniques for taking pictures have evolved to meet the demand.

Today, computer technology and highly advanced optical capabilities can give a glimpse of a world seen only in our imaginations. The current trend in much photography is to use exaggerated motion, bold colors, out-of-focus and grainy subjects, and off-center and skewed objects. And yet a need for traditional photography still exists. Those who have the required skills can make such photos work as an art form.

The advances in computer power have also influenced photography trends (Figure 1-12). If it a subject can be imagined, it can be created in the digital darkroom. For example, it is possible to make a believable image of a cow blowing bubbles. Computer technology makes the impossible possible.

Take a close look at any magazine. You'll see the trends just noted on the editorial pages and even more on the ad pages. Like writing, trends

FIGURE 1-11   Photographs from thirty years ago and beyond have lots of depth of field and a two-dimensional perspective. (*Courtesy USDA*)

FIGURE 1-12   Advances in computer graphics technology have made it possible to do what was once impossible. (*Courtesy USDA/ARS*)

come and go, but for now, the public responds favorably to grainy, out-of-focus, and skewed images.

THE FUTURE OF THE AGRICULTURAL COMMUNICATIONS INDUSTRY

It's hard to say what written or photographic trend will come next. Consumer preferences are fickle, and what works today may not work tomorrow. The one thing that is certain is that the need for outstanding agricultural communicators will continue to grow.

As the United States becomes more urbanized, people grow both physically and emotionally removed from farms. Their only connection to agriculture may be going to the grocery or the clothing store to pick up the products of agriculture. Because of their distance, the public has a diminished understanding of agriculture. Now instead of agricultural productivity, the public's primary concern is for food that is safe and economical.

With that fact so apparent, the need for agricultural communicators is apparent. From an industry-wide standpoint, agriculture needs people who can write, speak, and illustrate the positive points of agriculture. Training such people will bolster America's confidence in the most technologically advanced, economically priced, and safest food and fiber system in the world. Safe and economical agricultural goods are something that will never go out of style.

CONCLUSION

The agricultural communications specialist who has intimate knowledge of agriculture and can communicate effectively is important to the modern agricultural industry. The reason that agricultural communicators are important is that in an agriculturally ignorant society, the need for disseminating accurate information is crucial (Figure 1-13). An enlightened public is vital for agriculture to remain a strong industry in the consumer marketplace and in the halls of state and federal governments.

Agricultural communicators are trendsetters who can determine the look and feel of a publication or television show by the words or the visual effects they use. They are also trendsetters in the broader sense in that they can help shape positive consumer attitudes about the nation's food and fiber system.

To be a valuable communicator, an individual must have a concrete knowledge of both agriculture and communication trends. Recognizing and capitalizing on consumer trends increases the efficacy of written, spoken, and visual communications.

FIGURE 1-13   An enlightened public is important for agriculture to remain a strong industry. (*Courtesy Russell A. Graves*)

REVIEW QUESTIONS

Multiple Choice

1. An understanding of the processes by which food and fiber is brought from the farm to the consumer is
   a. knowing where milk comes from.
   b. agricultural literacy.
   c. agricultural illiteracy.
   d. knowing how a farm operates.

2. By developing a corps of agricultural communicators, the industry is _____ and _____ .
   a. fair, impartial
   b. biased, inaccurate
   c. unbiased, accurate
   d. skewed, unfair

3. What helps to put current trends and practices in perspective and illustrates a legacy of accomplishments?

   a. history
   b. a study of past trends
   c. neither a nor b
   d. both a and b

4. In 1819, what magazine was released in wide circulation?

   a. *Young Farmer*
   b. *Colonial Farmer*
   c. *American Farmer*
   d. *New Yankee Farmer*

5. What did Horace Greeley encourage people to do?

   a. Go west.
   b. Move to the country.
   c. Settle the western territories.
   d. Live off the land.

6. The American Agriculture Editors Association was organized in

   a. 1901.
   b. 1921.
   c. 1931.
   d. 1941.

7. What drives our popular culture?

   a. fads
   b. crazes
   c. trends
   d. none of the above

8. _____ are used to inform, persuade, and sell.

   a. Pictures
   b. Spin doctors
   c. Public information officers
   d. Words

9. What is the latest major means of disseminating news?

   a. newspapers
   b. radio
   c. television
   d. e-mail and Internet

10. Agriculture communicators can help shape positive consumer _____ about the nation's food and fiber system.

a. thoughts
b. ideas
c. attitudes
d. imagination

Fill in the Blank

1810	newspapers	weather
Morrill Act	600	art
Iowa State University	television stations	writing
1920	trend	literacy
vocal		

1. Congress established the nation's land grant university system when it passed the _____ _____.

2. Iowa State University began offering the bachelor of science degree in agricultural journalism in _____.

3. In 1990, there were over 102 agriculture _____ published and 5 _____ specializing in agriculture.

4. _____ offered the first class in agricultural journalism.

5. The first _____ _____ reports were broadcast over the radio from the University of Wisconsin.

6. In 1928, over _____ agricultural magazines were in publication.

7. A _____ is a pattern that people follow.

8. The first agricultural magazine was established in _____.

9. The _____ of _____ is a continually evolving process.

10. One of the most important tools that we can have in our intellectual arsenal is _____.

ENHANCEMENT ACTIVITIES

1. Using a variety of media such as television, magazines, newspapers, and the Internet, describe the current trends in the following areas.

   - automobiles
   - clothing
   - photographs
   - popular music
   - advertisements
   - food products

2. Research the history of agricultural communications and expand the time line in this chapter.

3. Find articles about the agricultural industry in popular press magazines and newspapers. Determine what percentage of the articles you have found portray agriculture in a positive light and what percentage in a negative light. Report your findings to the class.

WEB LINKS

To find out more about the topics discussed in this chapter, visit these Web sites.

- USDA: A history of American agriculture, 1776–1990
  <http://www.usda.gov/>
   (Search term: history)

- Visual arts trends
  <http://www.visualartstrends.com/>

- Consumer Trends Institute
  <http://www.trendsinstitute.com/>

- Agricultural Trends—Iowa State University
  <http://www.extension.iastate.edu/>

Alternatively, do your own search at: <http://www.alltheweb.com>
   (search terms: consumer trends, agricultural trends, visual trends,
      photography trends, agriculture communications)

BIBLIOGRAPHY

Burnet, Claron, and Tucker, Mark. (1994). *Writing for Agriculture: A New Approach Using Tested Ideas* (2nd ed.). Dubuque, Iowa: Kendall/Hunt.

Haygood, J., Hagins, S., Akers, C., and Keith, L. (2002). *Coverage of Media.* Paper presented at the National Agricultural Education Research Meeting. Las Vegas, NV.

GATHERING THE FACTS

OBJECTIVES

After completing this chapter, you should be able to:

- identify basic research techniques.
- identify effective interviewing techniques.
- practice fact-checking techniques.
- correctly attribute information.

KEY TERMS

alternative sources	interviewing	research
fact checking	plagiarism	

OVERVIEW

The importance of sound, unfettered research is one of the basic tenets of communicating in the agricultural industry. Effective communicators use good research and fact-gathering techniques. Solid facts do many things for a writer, including:

- solidifying credibility.
- gaining the confidence of readers.
- establishing the writer as a noteworthy communicator.

Whereas sound fact-finding techniques can bring you many rewards, sloppy techniques can have the opposite effect. Have you ever watched a television show or a movie and found a mistake in it, like a microphone that should have been hidden from a camera dangling conspicuously over the head of an unsuspecting actor? Although it may have been an honest mistake on the part of the directorial personnel, the production still loses a bit of credibility in your mind and makes the story a little less believable.

Solid, accurate research breeds confidence among your audience.

The same happens when you are writing content or photographing an event. You can lose credibility as a photographer if you create a news event just for the sake of a photograph and present it as the real thing without being truthful about its being a recreation. If you write a news article and don't bother to get all of the facts, your readers may not believe anything else you write ever again. A lack of confidence by your audience can ruin your career as an effective communicator.

In this chapter, you will learn how to get it right the first time. You will learn techniques that apply mostly to written communication but also to speaking to a group of people or shooting photos of a news event. You will also learn how to prepare for an interview and how to attribute information.

Above all else, this chapter will teach you how to gain and keep your credibility. As a communicator, that is your most precious commodity.

THE BASICS OF RESEARCH

Researching a topic can be a tedious process. Many good writers spend more of their time researching a topic than actually writing about it. That is because they have a desire to get all of the facts they can and then double-check those facts against other sources of information to ensure accuracy (Figure 2-1).

Good writers are also good researchers. They know what information is relevant to a story and what isn't. They have learned how to refine their techniques to the point at which they become efficient at finding and distilling information from a variety of sources.

The basics of **research** are simple: Choose a broad topic, develop some pertinent information about it, and refine the focus of the research to the point where the topic is narrow. The key is to be thorough in research. Thorough research will yield beneficial results to communications and help eliminate any mistakes in fact.

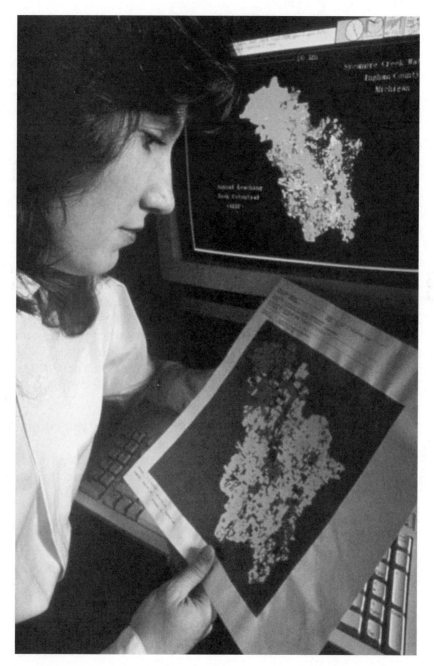

FIGURE 2-1   Research can be a tedious process. (*Courtesy USDA/ARS*)

The basics of research include finding out all you can about a particular subject.

Above all research always should be conducted. There is no one who knows everything about a particular subject. Your duty as a communicator is to find out all you can about a subject and then give the audience a thorough accounting.

Choose a Broad Topic

When presented with the challenge of producing some content, you usually start with a broad topic. In this way, you can view a whole subject and learn about it before deciding on what you'd like to write. In addition, the broad topic technique can yield many story ideas.

As an example, perhaps a magazine editor wants you to do a piece on a drought that has stricken the corn-producing states of the Midwest (Figure 2-2). Once you have researched the drought, you find that there are many subtopics within the broad topic—for example:

- the effect of the drought on corn prices
- how the drought is affecting farm families and communities
- the shortage of feed corn for cattle feedlots
- the trickle-down effect of corn prices on grocery prices
- the drinking water shortage that the drought is influencing

Any of these topics could be developed into a viable news story with additional research. By choosing a broad topic, you give yourself options as to which direction to take a story.

The same concept of choosing a broad topic also applies to preparing an oral presentation. Suppose a civic club wants you to talk about the local FFA chapter. To give an effective speech, explore all of the possible topics that you could speak about, and narrow it to one.

Narrow Your Focus

Once you have a solid overview of the subject by conducting sound and thorough research, it's time to narrow the focus. Let's use the example of talking about the local FFA chapter at a civic club meeting. The FFA is a huge organization with many different facets. Where do you start?

FIGURE 2-2 Agricultural subjects, such as corn, are often broad topics full of subtopics, such as the effects of drought on prices. (*Courtesy Russell A. Graves*)

First, identify what your chapter participates in and divide those activities into categories: Career Development Event(s) (CDEs), mentoring programs, Supervised Agricultural Experience(s) (SAEs), and others (Figure 2-3). Once you have narrowed the focus, you can get to work on preparing the speech.

The reason for narrowing focus is that your venue is often limited. Whether it is a book, a magazine or newspaper article, or a stage in front of a coliseum filled with 10,000 people, the space you have to write or the time you have to speak is always limited.

Could you write a 2,000-word article on the history of ranching in the United States? You could, but the piece would be so boiled down to the basic facts that it probably wouldn't be very interesting because it lacked substance. You could write a whole book on the subject and still not cover every detail of the history of cattle ranching (Figure 2-4).

Instead, focus on an aspect of the subject, research it thoroughly, and present it in a way that people can connect with the material they are reading on an emotional or a cognitive level. You may have noticed that books about ranching typically feature stories about an individual or family and how they conquered the elements in a classic story of survival. When you

FIGURE 2-3   The FFA is a broad subject. You can narrow it by talking just about SAEs, such as care of pigs. (*Courtesy Russell A. Graves*)

put a human element into a story, people can connect with the piece. If your approach is from a broad standpoint, then you can't boil it down to its most interesting elements.

Although the points about narrowing subjects have been presented as pieces that were assigned by an editor or requested by a civic club leader, you should practice the same techniques with a self-assignment: first developing a broad topic, then getting an overview of that topic, and then narrowing the focus. Once the focus is narrowed, you can move on with research.

Hit the Books

The research process may be the most important step in communicating information and the most time-consuming. When working on a piece, many writers spend weeks and even months doing the background work and uncovering every possible detail about a story before actually sitting down and writing. Research is part of the writing, speaking, and photographic process.

Good research using good sources yields useful information so that you can report the facts accurately and thoroughly. Research materials

FIGURE 2-4   Writing a 2,000-word story about a subject as broad as ranching would be a difficult task without narrowing your topic.

come in many forms: personal interviews, books (Figure 2-5), magazines, television shows, and Internet resources. All of these sources can be valuable in helping you gather facts.

Today, however, researchers have to be particularly careful. With the proliferation and ease of printing and Web-based technologies and their ease of use, *anyone can publish anything*. Material doesn't have to be factually correct to be published. Errors in published work may be a result of

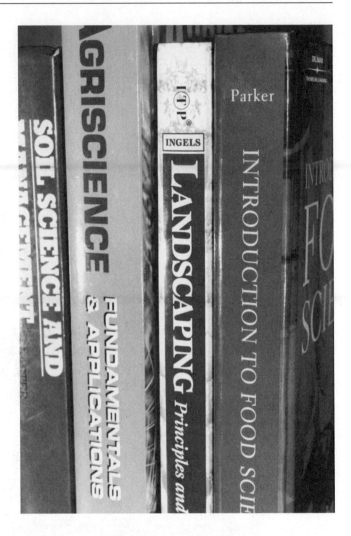

FIGURE 2-5   Good sources of research are important for thorough and accurate reporting. (*Courtesy Russell A. Graves*)

poor proofreading, incorrect source material, or even a malicious attempt to spread misinformation.

Therefore, good communicators look at sources with a bit of a cynical eye and don't take anything at face value until it has been verified by multiple sources. That rule applies whether your source is a person, a book, a magazine, or a Web site.

As a communicator, when you gain credibility in the accuracy of your reporting, you too could become a source for someone else's work. If your work is solid, chances are that people will come back to your work for additional information.

Many sources of information are credible because they have a long-term reputation for their accuracy. Many magazines including *Time*, *Progressive Farmer*, and *Newsweek* have developed reputations as reliable sources of information. When researching, always try to seek out first those publications with a solid reputation for truthfulness.

INTERVIEWING TECHNIQUES

Interviewing is an art. Oprah Winfrey, Barbara Walters, Bill O'Reilly, and Larry King are a few of the premier interviewers on television, and they are rewarded handsomely for their skill in talking to people.

The reason for interviewing someone to gain information is to put a human face on a subject. Recall the example when a magazine editor asked you to write about drought in the Midwest (Figure 2-6). Let's say that once

FIGURE 2-6   Interviewing can yield information—for example, the effect that drought is having on a farmer's crops.

you narrowed your topic, you decided that it would be interesting to write a piece on how the drought is affecting farm families. Without interviews, you could not gauge the true impact of a drought on someone who works the land. Interviews help people connect emotionally with the plight of farmers.

There are some ground rules to abide by when interviewing others (Figure 2-7). A good interviewer knows those rules and applies them always. They include:

- Prearrange interviews.
- Get background information on the subject.
- Ask questions relevant to the subject of the interview.
- Be respectful of someone else's opinion.
- Be a good listener.
- Thank the interviewee for his or her time.
- Follow up with your interviewee to clarify any unclear information.

These basic rules of interviewing may look familiar—they are also basic principles of getting along with people in your everyday life.

Where television personalities differ from most other interviewers is that they have developed a credibility that allows them to interview almost anyone on almost any topic and allow their guests the comfort to tell all on national television.

With a little practice, you can be a good interviewer too. By following some basic principles, you can get the information needed to put a personal touch on any story and give depth to your coverage of a subject.

- Be respectful of someone else's opinion.

- Be a good listener.

- Get background information on the subject.

- Ask questions relevant to the subject of the interview.

- Prearrange interviews.

- Follow up with your interviewee to clarify any unclear information.

- Thank the interviewee for his or her time.

FIGURE 2-7 The rules.

Do Some Background Checking

An interview actually begins well before you meet the person. In many events, such as fast-breaking news stories, you can't do any background checking on the interview source, so you have to interview someone who was present at the event. For feature stories, though, you often have several months to find and cultivate sources for a report (Figure 2-8). In that case, do some background checking.

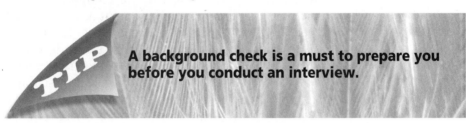

TIP A background check is a must to prepare you before you conduct an interview.

Corral Extra Profits With Quality Working Pens

By Russell Graves

s cattle raisers, we all know e importance of having iality equipment in our op- ation. Let's face it, every ar we spend millions of llars collectively on the ings that we use to help us ong in our never ending irsuit to make ends meet. ings that, for this particular ample, we'll term hard- are.

Now, what I consider e hardware of our industry e those items that we abso- tely have to have to create ir product; that being qual-

spent on equipment, produc- ers are constantly looking for ways to cut their costs when- ever they get the chance. All to often, however, the cuts made are in the area of cor- rals and working pens. Like with cattle, if you invest in sub-quality or an inadequate product, you are setting your- self up for a loss.

Now, you may be wondering how working pens can adversely affect your bot- tom line. After all, you've bought the best cows money can buy, your AI representa-

out of them. What this ty of setup results in is an creased amount of cussi and severe frustration on part of the cowman. In tu cattle may end up not bei dewormed according schedule, calves may weaned off late, or culls m hang around in the pastι too long.

In a study, conduc by the Georgia Cooperat Extension Service, it sta that, "...If there are any wiι fall profits to be made in cattle business, then it is dc

FIGURE 2-8 For feature stories, you often have many months to find and cultivate sources for reporting. (*Courtesy Russell A. Graves*)

The background check will do many things for you.

1. It helps you determine the relevance of your source to the piece you are writing. If you are planning to interview a college professor on bioengineered cotton varieties, make sure that the person you have identified to interview has some experience in this field. A background check will assist you to that end.
2. It will help you determine the credibility of a source. An interviewee's credibility is as important as any printed material. If you were to interview someone who is less than credible, your reputation as a factual reporter could be put in jeopardy.
3. It acts as a discovery process. By surveying a person's body of work, you can decide ahead of time the right questions to ask. Then you can get to the point of an interview quickly and gather lots of relevant information. A background check is part of the research process, just as looking through books is.
4. It can help you determine the interests of the individual you are interviewing. Suppose you are interviewing a cooperative extension agent and discover in your research that she is an avid angler. To put her at ease, you could chat about fishing before you begin the actual interview.

Come Up with Some Predetermined Questions

Professional interviewers always have predetermined questions so that they can guide their interview subject smoothly from one point to the next. Often, the interviewee is given the opportunity to preview these questions so that he or she might avoid any awkward moments by not knowing the answer to a particular question.

Before you step into an interview, always think of some predetermined subjects.

When coming up with a list of questions, always start with questions that give a general overview of the subject. After that, get into the specific points of a subject. Save any sensitive questions for the end.

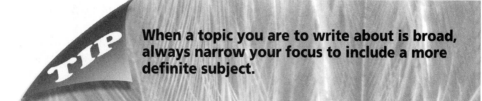

An interview with a beef biotechnologist might use questions that unfold like this:

1. Tell me about the work you are doing with beef cattle and biotechnology.
2. In terms of the biotechnology industry, how does your work compare to other research that is being done?
3. What breeds of cattle are you focusing on in your research?
4. Specifically, what does your research entail?
5. What are some genetic techniques you employ in your research?
6. Does the research involve any danger to the animals?
7. What would you say to those who would claim that altering an animal's genetic makeup is unethical?
8. Do you feel that the technology you are using may be perceived as unethical?

You can see that these questions unfold in a logical manner. They start by asking the interviewee for a broad overview of the work she is doing. Next, they steer the interviewee in a direction that delves into the details of the research. In the last three questions, the interviewer queries the interviewee on more sensitive issues of her research. The tough questions are usually presented last so that if the interviewee balks, the core information has already been obtained.

Being prepared for an interview shows your professionalism. If you have the correct background information and know the right questions to ask, you'll be surprised at how candid subjects will be. And it is the ability to bring out the candidness in an interview that sets good interviewers apart from bad interviewers.

Ad-Libbing During an Interview

Sometimes you will need to digress from the prepared questions that you started the interview with. Whether it is for clarification purposes or you think of another question to pose to the interviewee, the ability to ad-lib during an interview is important. Although technically, *ad-lib* means to improvise as you go along, you can be prepared to ad-lib if you know the background information on a subject well.

A good interviewer also has a set of backup questions that relate to the prepared questions and are asked only in response to how an interviewee answers a question (Figure 2-9). For example, in the previous section, question 8 asked, "Do you feel that the technology you are using is unethical?"

FIGURE 2-9   A good interviewer often has a set of backup questions prepared. (*Courtesy USDA/ARS*)

If the interviewee answers simply yes or no, you can ad-lib by asking, "You said that you think the technology you are using is ethical. Why is that your opinion?" If the interviewee answered no, you could still pose the same question.

Ad-libbing helps you clarify what the interview wants to go on the record. If you don't ad-lib to clarify information, then you run the risk of misrepresenting what the subject of the interview said or distorting that person's view on an issue. Either way, your credibility could be damaged.

CHECKING FOR FACTS

Fact checking should be a constant process that you do with the same seriousness and diligence that you pursue when writing an article or interviewing a subject.

Your duty as a speaker, photographer, or writer is to bring factual information to your audience. Perhaps you took a picture of a golden field of wheat for the cover of an agricultural periodical, but when you sent the slide to the magazine, you captioned it as being oats. If the periodical runs the image and identifies it as a field of oats, it is likely to hear from perceptive readers who will point out the error. With a little fact checking, you would have known that the photograph was of wheat (Figure 2-10).

A mistake like a mislabeled image can be an embarrassment for a magazine and cost you business in the end. Remember that whatever you say, whether in words or pictures, people are paying attention and will notice your mistakes. Eliminate the mistakes with sound fact checking.

Finding Alternative Sources

As a fact checker, your challenge is to think creatively when coming up with **alternative sources** of information, that is, looking at lots of information from magazines and books to television transcripts, Web sites, and previously published interviews. Finding alternative sources takes time, but your audience and your credibility deserve the time it will take.

TIP

Thinking creatively is an important skill for agricultural communicators.

FIGURE 2-10 By fact checking you can determine whether this is an image of wheat or oats. (*Courtesy Russell A. Graves*)

Second Opinions

Someone once said, "I use not only the brains I have but all that I can borrow." When fact checking, those words are valuable advice.

Even if you are well versed on a subject, there is always someone who knows more. It is always valuable to have other people read your material to see if they notice any errors of fact. In addition, they may be able to point out where clarification is needed.

Often, when agricultural journalists write a piece and quote someone, they will let the person quoted read the draft to confirm if the quotation is accurate. For other second opinions, try asking teachers, college instructors, cooperative extension agents, scientists, and other experts in various fields for their opinion on the facts presented in your work.

ATTRIBUTING INFORMATION

A magazine editor once opined, "Everything is attributable." In other words, information doesn't materialize out of thin air; someone divulges it. When working on a speech or article, you bolster your credibility by using phrases like "according to . . ."

Attributing information is not always enough. If you plan on using information verbatim, it is a good idea to contact the author of the information and obtain, in writing, permission to republish his or her work.

When you attribute information, you show the audience that you have done your research on the topic and gotten others' points of view on the subject. By doing so, you again help bolster your credibility as a speaker or writer. In addition, by attributing information, you show a measure of humbleness.

Many inexperienced writers try to piece together articles that attribute no information—they present fact after fact without divulging a source. What they are left with is a piece that looks as if all of the ideas presented belong to the author.

Attributing information can also have another benefit. Suppose you are reporting a controversial story. If you don't attribute, readers may get the sense that you side with the controversy. By attributing information to someone else, you shield yourself from any negative backlash that may ensue, preserving your respect and credibility as a communicator.

Remember that all information is attributable. If you put a specific fact in a speech or a magazine article, be sure to identify the source. If you attribute, you allow the readers to probe the subject if they choose, and you leave a recorded paper trail of information that allows researchers to use your information as a trusted resource for their articles.

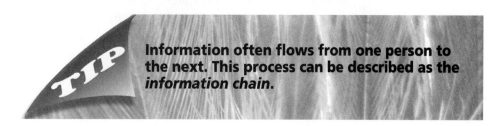

Information often flows from one person to the next. This process can be described as the *information chain*.

Citing Information

When you write a research paper, you typically list the publications you used at the end of the paper.

You probably learned in your high school language arts classes that there are many ways for using citations in a paper. Among the most popular are citations formats following the Modern Language Association's (MLA) guidelines. MLA guidelines are pretty straightforward and follow different styles according to whether the citation is a book, magazine article, website, or other published material.

MLA style is among the most popular although various citation styles are outlined by the American Psychological Association (APA) and the Chicago Manual of Style. Different publications desire varying forms that they outline to perspective authors. Often, these citations look something like this:

Graves, Russell. *Communicating in the Agricultural Industry*. Clifton Park, N.Y.: Delmar Learning, 2004.

These citations guide readers to publications featuring work similar to yours should they desire more information. The references also allow readers to fact-check your work should they need to verify the information you have presented.

With magazine and newspaper articles, as well as speeches, you don't have the luxury of adding a long list of sources at the end of your written

Gillian, Rick. "*The Blacktailed Prairie Dog.*" Ii
      terly, date unknown.
Frisina, Michael, and Jine Marini. "Wildlife an
      ments of the Grassland Ecosystems." F
Glass, Anthony. *Journal of an Indian Trader: A
      Texas Trading Frontier*, 1790–1810. Ed.
      Station: Texas A&M University Press, 1
Haley, J. Evetts. *Charles Goodnight, Cowmar
      University of Oklahoma Press, 1936.
Hansen, Richard, and Ilyse Gold. "Blacktail P
      Cottontails and Cattle Trophic Relations
      Range." *Journal of Range Management*

FIGURE 2-11   Bibliographical citations point readers to the sources of your information.

piece or speech. Instead, your citations must be in the form of the words you use. Very little information that is around is purely original in nature but points of view and thought are. Therefore, if you are directly quoting someone else's interpretation of an issue, then an in-sentence citation should be used—for example, "According to *Time Magazine*, the incidence of hoof and mouth disease in the United Kingdom is declining, and the epidemic is under control."

It doesn't matter if you use a citation of this manner, or any other type, in the written or spoken word. By citing sources of information, you build credibility and avoid the trap of plagiarism.

AVOIDING PLAGIARISM

It is easy to copy information and present it as your own. **Plagiarism** is a piece of writing copied from another source that is presented as the author's own. It is a sad fact that plagiarism occurs constantly in high schools and colleges all over the country. With the ease of copying and pasting information using a computer, the lure to take someone's words and use them as your own is strong. And if no sources are provided, the chances are that most people would never know where the information came from.

Plagiarism is a trap that should be avoided.

In fact, plagiarism is illegal. Based on the extent of the plagiarized act, the fine can run into the thousands of dollars for each offense. Worse, plagiarism can cost you a good reputation.

Remember to cite and attribute all information. Moreover, when you present information, always reword it into your own thoughts. Make the idea a manifestation of your interpretation and not a facsimile of someone else's work.

ETHICAL CONSIDERATIONS

In this chapter, we've discussed the importance and need to research a topic carefully and accurately. We have also touched on the importance of attributing information to the proper sources, as well as avoiding the trap of plagiarism at all costs.

As an agricultural communicator, you must adhere to the highest code of ethics so that your work will be respected by those who read or view it. A lifetime of solid, trustworthy work can be ruined by a single instance of dishonesty.

A definition of ethics is "the science of human duty; the body of rules of duty drawn from this science; a particular system of principles and rules concerning duty, whether true or false; rules of practice in respect to a single class of human actions; as, political or social ethics; medical ethics." From this definition, it is clear that ethics are actions that are governed by the duty involved.

In the agricultural communications field, the need for the most stringent of ethics is acute. Agriculture is in a very public fishbowl. Everyone needs the products the agricultural industry produces, yet the public is quick to criticize the actions and practices of the industry.

Although your allegiance may be toward agriculture, you still have a duty to yourself and whomever you represent to remain staunchly accurate. With that in mind, here are a few ethical considerations to think about:

Do not let personal feeling cloud your judgment   At some point, you may have to report on something or someone you do not like. Even harder, you might have to report unflattering news on something that you do support. Never let your personal feelings get in the way of the absolute truth. If you cannot separate the personal from the professional, tell your superior, and get reassigned to another project.

Respect others' opinions   As an agricultural communicator, you might have to interview someone who opposes an agricultural practice that benefits farmers in the area. In order to give fair and balanced coverage of a topic, respect other people's opinions on a subject even if they are the complete opposite of your own opinions.

Treat other people as you would like to be treated   Would you like to be goaded into answering a question you feel uncomfortable or unsure about? This ethical consideration is one your parents probably taught you when you were young. It applies in the professional world as well.

Always strive for accuracy and truth   Much of this chapter has dealt with the need for you to be accurate in all you communicate. Even if the news you are asked to report casts an unfavorable light on a particular agricultural topic, you should nevertheless strive to be accurate in your reporting. With accuracy comes truth. With truth comes integrity.

Maintain your integrity   Always do the right thing in all situations in order to be a credit to yourself, your name, and the people you represent.

CONCLUSION

This chapter has looked at ways to research information. From consulting books and magazines to conducting interviews, good research is the cornerstone of good writing. With research and writing come responsibilities, especially attributing information to the proper sources. Attribution is one of the foremost ways to let people know that you are not a know-it-all and to preserve your reputation as a reputable communicator. Attribution can also shield you from any criticism should you be reporting on a controversial issue.

Research can make or break the career of a communicator. If you cherish your reputation for credibility and reliability, you'll conduct good research and avoid plagiarism at all costs. Remember that plagiarism can cost you more than money—it can cost you your integrity and your reputation.

REVIEW QUESTIONS

Multiple Choice

1. Solid facts do many things for a writer, including

   a. solidifying your credibility.
   b. gaining the confidence of your readers.
   c. establishing yourself as a worthwhile communicator.
   d. All of the above

2. A broad topic technique can

   a. lengthen the time spent doing research.
   b. help increase your knowledge on topic.
   c. yield many story ideas.
   d. decrease the time spent doing research.

3. The reason for narrowing your focus is that

   a. too much information can be confusing.
   b. it is easier to write about a narrow subject.
   c. most people are narrow-minded.
   d. your venue is often limited by time or space.

4. Good research coupled with good sources yields

    a. a lot of information that must be sorted through.
    b. useful information.
    c. knowledge about the subject to be reported on.
    d. Both b and c

5. If your work is solid, the chances are that

    a. people will come back to your work for useful information.
    b. it will be harder for people to get through it.
    c. it will take longer to write the piece.
    d. None of the above

6. The reason for interviewing someone to gain information is

    a. to meet new and exciting people.
    b. a way of advertising your work.
    c. to show that you are serious about the subject.
    d. to put a human face on the subject.

7. A background check can be a

    a. way of finding a criminal history of the interviewee.
    b. way of determining security clearance.
    c. discovery process.
    d. All of the above

8. Before performing an interview, it is important to have

    a. showered.
    b. predetermined your questions.
    c. gone to the restroom.
    d. make up the questions as you go.

9. The sequence of questioning in an interview should be

    a. sensitive, specific, and general questions.
    b. specific, general, and sensitive questions.
    c. general, sensitive, and specific questions,
    d. general, specific, and sensitive questions,

10. Sound fact checking

    a. lets your boss know that you know what you are doing.
    b. eliminates mistakes.
    c. clarifies what should go on the record.
    d. is unnecessary.

Fill in the Blank

background
cornerstone
cynical
emotional
first
guide
most
narrow
options
pertinent
photographic

plagiarism
second opinion
shield
truthfulness
relevance
manifestation
multiple
connect
avoid
writing

similar
cognitive
thorough
broad
focus
speaking
interpretation
factual
attribute
credibility

1. The basics of research are simple. First, choose a broad topic and develop _____ information. Then refine the _____ to the point that you have a _____ topic. The key is to be _____ in your research.

2. By choosing a _____ topic, you give yourself _____ as to which direction you would like to take the story.

3. Focus on an aspect of the subject, and present it in a way that people can _____ with the material they are reading on either a(n) _____ or a(n) _____ level.

4. The research process may be the _____ important step in communicating information.

5. Research is part of the _____, _____, and _____ process.

6. A good communicator looks at sources with a bit of a _____ eye and does not take anything at face value until it has been verified by _____ sources.

7. When researching, always try to seek out _____ those publications with a solid reputation for _____.

8. A background check can help you determine the _____ of your source to the piece you are writing and also the _____ of a source.

9. Your duty as a speaker, photographer, or writer is to bring _____ information to your audience.

10. When someone else reads your material to see if he or she notices any errors in fact, that person is supplying a _____ _____.

11. When you _____ information, you show your audience that you have done your research on the topic and gotten others' points of view on the subject.

12. By attributing information to someone else, you _____ your-self from any negative backlash that may ensue if the subject is controversial.

13. Citations _____ the reader to publications featuring work _____ to yours.

14. If you are directly quoting someone else's _____ of an issue, use an in-sentence citation.

15. By citing information, you _____ the trap of plagiarism.

16. Make the idea a _____ of your interpretation and not a _____ of someone else's work.

17. Good research is the _____ of good writing.

18. _____ is illegal.

19. If you _____ sources, researchers can use your information as a trusted source.

20. _____ will always ensure credibility as a communicator.

ENHANCEMENT ACTIVITIES

1. Choose one of the topics listed and narrow its focus. Then, using the research techniques described in this chapter, write a paper (of the length prescribed by the teacher) on the subject.
   - biotechnology
   - bioterrorism
   - livestock genetics
   - modern crop production practices
   - environmental concerns of agriculture
   - precision farming

2. Interview a classmate about his or her life history. Before the interview, come up with a list of questions you'd like to know about the person. After the interview, write a short paper about the student you interviewed.

3. Interview a local farmer, rancher, agricultural scientist, or agribusiness leader about his or her job and the impact that this person is having on local agriculture.

4. Write an article about your local agricultural economy, and attribute at least five pieces of information that you gathered.

WEB LINKS

To find out more about the topics discussed in this chapter, visit these Web sites.

For conducting research on the Internet:
<http://library.albany.edu>

A student's guide to research with the WWW
<http://www.slu.edu>
  (Search terms: english department, research)

Using Modern Language Association (MLA) format
<http://owl.english.purdue.edu>
  (Search term: research handouts)

A guide for writing research papers based on Modern Language
    Association (MLA) documentation
<http://webster.commnet.edu>

Tips for conducting interviews
<http://www.netwrite-publish.com>
  (Search term: interview tips)

Or do your own search at <http://www.alltheweb.com>
  (Search terms: interviewing techniques, researching, plagiarism,
    research papers)

BIBLIOGRAPHY

Burnet, Claron, and Tucker, Mark. (1994). *Writing for Agriculture: A New Approach Using Tested Ideas* (2nd ed.). Dubuque, Iowa: Kendall/Hunt.
Holt, Rinehart and Winston. (1991). *Writing the Research Report*. Austin, Tex.: Author.

CHAPTER 3

WRITING FOR AGRICULTURE

OBJECTIVES

After completing this chapter, you should be able to:

- understand the basics of journalism.
- write an effective press release.
- write a feature story.
- edit written work effectively.
- appreciate the need for effective communications.

KEY TERMS

editorializing	lead	news determinants
feature story	news components	news release

OVERVIEW

Writing is an elemental form of communication. For thousands of years, people have used the written word to communicate messages, record history, and create fantasy (Figure 3-1).

Early on, written word was in the form of pictures painted on cave walls. As humankind's knowledge base began to expand, symbols took the place of pictures to convey knowledge. As the symbols became more refined, they evolved into words.

Before the printing press was invented in the fifteenth century, books were painstakingly scribed by hand, one page at a time. Since then, the efficiency with which knowledge is shared through written communications

FIGURE 3-1   From the earliest times, humans have used
symbols to communicate. (*Courtesy Russell A. Graves*)

has expanded exponentially. Books, and the knowledge they contained,
could be duplicated and shared on a huge scale. As the years passed, the
printing process became increasingly refined. Now electronic printing and
publishing is the rule. With computer technology, you can see what your
document will look like before you go through the expense of printing it.

The power to mass-produce documents is at the fingertips of literally
everyone. And with that power comes responsibility. Agricultural communicators have a responsibility to represent their employer and the industry
in the best light possible. That responsibility begins with the ability to
communicate effectively by using the written word.

In this chapter, you will learn the basics of journalism as they apply to effective agricultural communications. To this end, you will learn the elements of an effective **news release**, as well as the things that make a **feature story** a good one. You will also learn how to effectively edit your work as well as the work of others.

This chapter highlights the importance of conveying your thoughts and information through the written word. Writing is a powerful medium for communication. People who understand that use their ability to write to position themselves so that they, as well as the people they represent, will benefit.

THE NEED FOR EFFECTIVE WRITTEN COMMUNICATION

E-mail and the Internet are great for communicating messages and ideas. In a way, electronic communicating has also revived the art of letter writing: more people now take the time to write messages to others than perhaps did before the proliferation of e-mail. As such, e-mail and the Internet are a boon for people who cherish the written word.

Effective written communication conveys the author's message in an efficient, straightforward, and easy-to-understand manner (Figure 3-2). Effectively written messages are powerful persuasion tools and put the writer in a favorable light.

Electronic communication has done much to revive writing, but it has also introduced some aspects to the medium that could be described as detrimental. Developed to aid in efficiency at communicating a message, abbreviations of common phrases often creep into other types of writing. For effective agricultural communications, these abbreviations have no place whatsoever. Here are a few examples of common Internet and e-mail abbreviations.

- IMHO means "In my humble opinion."
- AFAIK means "As far as I know."
- U means "you."
- C means "see."
- L8R means "later."

Unconventional abbreviations often lead to confusion.

FIGURE 3-2   With the proliferation of the written word, there is a strong need for effective communications. (*Courtesy of USDA/NRCS*)

Effective professional communications, whether for the agricultural or any other type of industry, never use these abbreviations. Instead, you should learn to express your thought fully without excessive wordiness, abbreviations, or unconventional means of communicating.

The need for effective writing is indeed strong. No matter where you look, everyone uses writing to communicate. Think about how billboards, street signs, businesses, and municipalities use writing to communicate. Then look at how magazines, newspapers, television, and all sorts of advertisements use writing as a basis for getting their message across (Figure 3-3). Suppose the message that an advertiser was trying to convey was not clear or had mistakes in it. That would probably shake your confidence in the company.

FIGURE 3-3   Advertising uses words to communicate messages about everyday products. (*Courtesy of USDA/NRCS*)

Companies know that a poorly crafted message can reflect back on them, and they endeavor mightily to avoid such problems.

Just like large corporations, you too can make efforts to see that your message is clear to all who read it. It is not hard to do at all.

Spelling

Not everyone is a perfect speller. (If they were, there would be no point in having spelling bee competitions.) The good news is that good writers do not have to be perfect spellers. They only have to be able to recognize when a word is misspelled and take action to correct it.

As great as computer software is, it should not be used as a crutch in order to find misspelled words. Most word processing software programs point out misspelled words, but they do not catch words with transposed letters that spell another common word. A simple example is the word *two*. Using the same letters, it is easy to spell *tow*, and the spell-check program will not pick up the mistake.

Aside from the computer, you should also keep a dictionary handy as a supplement (Figure 3-4). A dictionary cannot confuse words and works well to provide the definitions and context of words.

Remember that words are a two-edged sword. Correctly spelled words can yield power and efficacy to your communications; misspelled words sow the seeds of doubt and instill a lack of confidence in your readers.

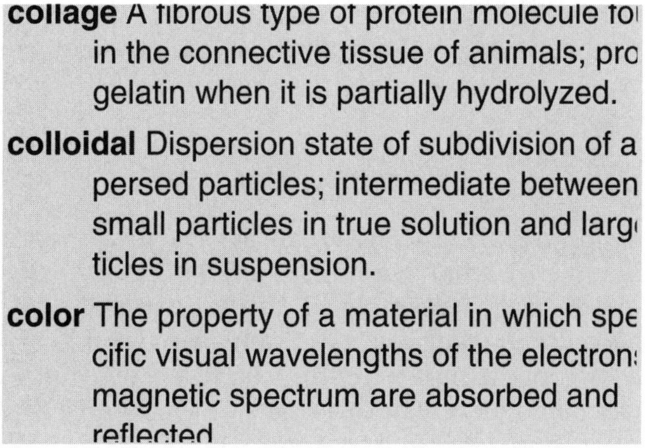

collage A fibrous type of protein molecule fou in the connective tissue of animals; pro gelatin when it is partially hydrolyzed.

colloidal Dispersion state of subdivision of a persed particles; intermediate between small particles in true solution and larg ticles in suspension.

color The property of a material in which spe cific visual wavelengths of the electron: magnetic spectrum are absorbed and reflected

FIGURE 3-4   Aside from the one on the computer, you should also have a traditional dictionary as a handy reference.

Always remember the spelling lessons that you learned in elementary school—rules like "I before E except after C." Remember too that not all words are spelled phonetically. If they were, *phonics* would be spelled *fonix*.

Grammar

If you view correct spelling like gasoline in a car, then grammar is the engine that makes it run. Proper grammar is more important than correct spelling. Analyze the following two sentences.

Merit lights springs temple, east mile loop exit?
A sexshon of land consis of 640 ackres.

Which sentence makes more sense? Does the first or the second sentence convey a message? The first sentence has no misspelled words, while the second one has three. The main difference between the two is that the second sentence has a grammatical structure. The words used fit together with a semblance of context. Although sloppily misspelled, the second sentence does convey a message, while the first is simply a collection of words.

Throughout your schooling, teachers have coached you on grammar. Think of agricultural journalism as the game you get to play after being coached. Writing stories and news releases is putting the lessons you learned in the classroom into action.

The more effectively you put those lessons in action, the better your writing will be and the more convincing voice you will have. When teamed with correct spelling, grammar is a powerful tool that can last an eternity.

Writing for an Audience

When you are preparing to write, always do an audience assessment because your audience will have a direct influence on the type of writing that you will produce. For example, if you are writing a paper on research that has been conducted, you will use highly technical language and cite references to supporting literature. If you are writing a piece about the local FFA chapter's success in this year's CDE contests, then the words you use will be far more general in nature.

Most popular magazine articles are written on a junior high reading level (Figure 3-5). The reason is that magazines want to appeal to as many people as possible. Because reading skills vary considerably, many maga-

FIGURE 3-5   Magazine articles are often written on a junior high level to appeal to a broader audience. (*Courtesy of USDA/NRCS*)

zines know that by writing to a lower reading level, they can appeal to a broader range of buyers and subscribers.

To write for a particular audience takes practice. Usually, writing for a lower grade level is easy. Writing for advanced readers takes practice and challenges your prowess.

JOURNALISM BASICS

According to the Texas Tech University's Department of Agriculture Communications' *Agricultural Communications CDE handbook*, the basics of journalism can really be described as the ABCs of journalism: accuracy, brevity, and clarity (Figure 3-6).

Accuracy is necessary in any credible communications, written or spoken. Unless the work is clearly presented as fictional, there is no excuse for material that falls short of being 100 percent correct. Accurate writing begins with good, solid research. As discussed in Chapter 2, facts on a story should be checked, double-checked, and then checked again. The fact

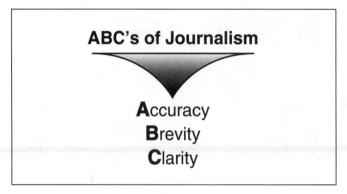

FIGURE 3-6   ABCs of Journalism: Accuracy, Brevity, Clarity.

checking should always come from reliable sources. Reporting falsities perpetuates misinformation and further clouds the truth.

Brevity relates to the conciseness of an article. In newspapers, magazines, and, to a lesser degree, Web sites, space is a premium. Most magazines look for articles that are 1,000 to 2,000 words, while many newspapers rarely use articles much more than 500 words. Web sites are unique in that on the home page and other key parts of the site, brevity must rule. When users log on to a Web site, they want information quickly. If you can catch a Web browser's attention without wasting that person's time, then you can direct him or her to areas that contain longer content.

The last part of the ABCs of journalism is *clarity*—the ability of the written word to get the point of the article across to readers. Does what you just wrote make sense? Can a reader discern the meaning of the piece in the words that you have been allotted? These are questions to consider whenever you write a piece of work intended for someone else to read. One of the best ways to ensure clarity is to have someone else versed in spelling and grammar read a piece and provide an objective critique of the work. If writing is clear to someone else, chances are that it will be clear to nearly everyone who reads it.

Clarity also demands sticking to the facts of a piece and not editorializing. **Editorializing** means that the writer has infused his or her opinion into a piece of written work. Although editorial writing is appropriate for some venues, most news stories should avoid opinion and stick to the facts. Facts allow readers to make their own decisions and form their own opinions about the subject of a piece and lend clarity to writing.

Writing that closely follows the ABCs of journalism is a powerful tool that can persuade and inform audiences. As with any other discipline, a writer's ability to deliver a professional product will increase the demand for his or her services.

PREPARING NEWS RELEASES

News releases are tools that corporations, small businesses, educational institutions, and other organizations use to get the story out on their breaking news and events.

Effective news releases usually follow an established format. They begin with a strong **lead**. The lead generates attention for the story. A good story headline gets people to stop and pay attention to a piece, and a good lead paragraph brings a reader into the story.

A good lead starts with the basics: who, what, where, when, why, and how (Figure 3-7). In the first paragraph of the piece, all of the pertinent facts are presented. That way, a reader can decide quickly if the story is something important or useful to read. Subsequent paragraphs of a news story provide supporting information that relates to the lead paragraph.

This style of writing is often referred to as the inverted pyramid (Figure 3-8). The inverted pyramid style of writing puts the most important facts of a story at the beginning of an article and subsequently puts secondary and tertiary facts later in the piece.

Using the inverted pyramid writing style ensures a continuity of facts and a naturally flowing means of providing information. It also serves as a tool for organizing your information when preparing to write a news article.

The following news release was distributed by Iowa State University's Cooperative Extension Service.

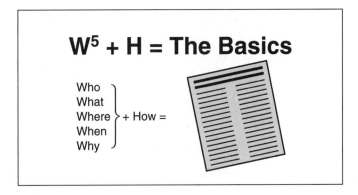

FIGURE 3-7 W^5 + H = The Basics.

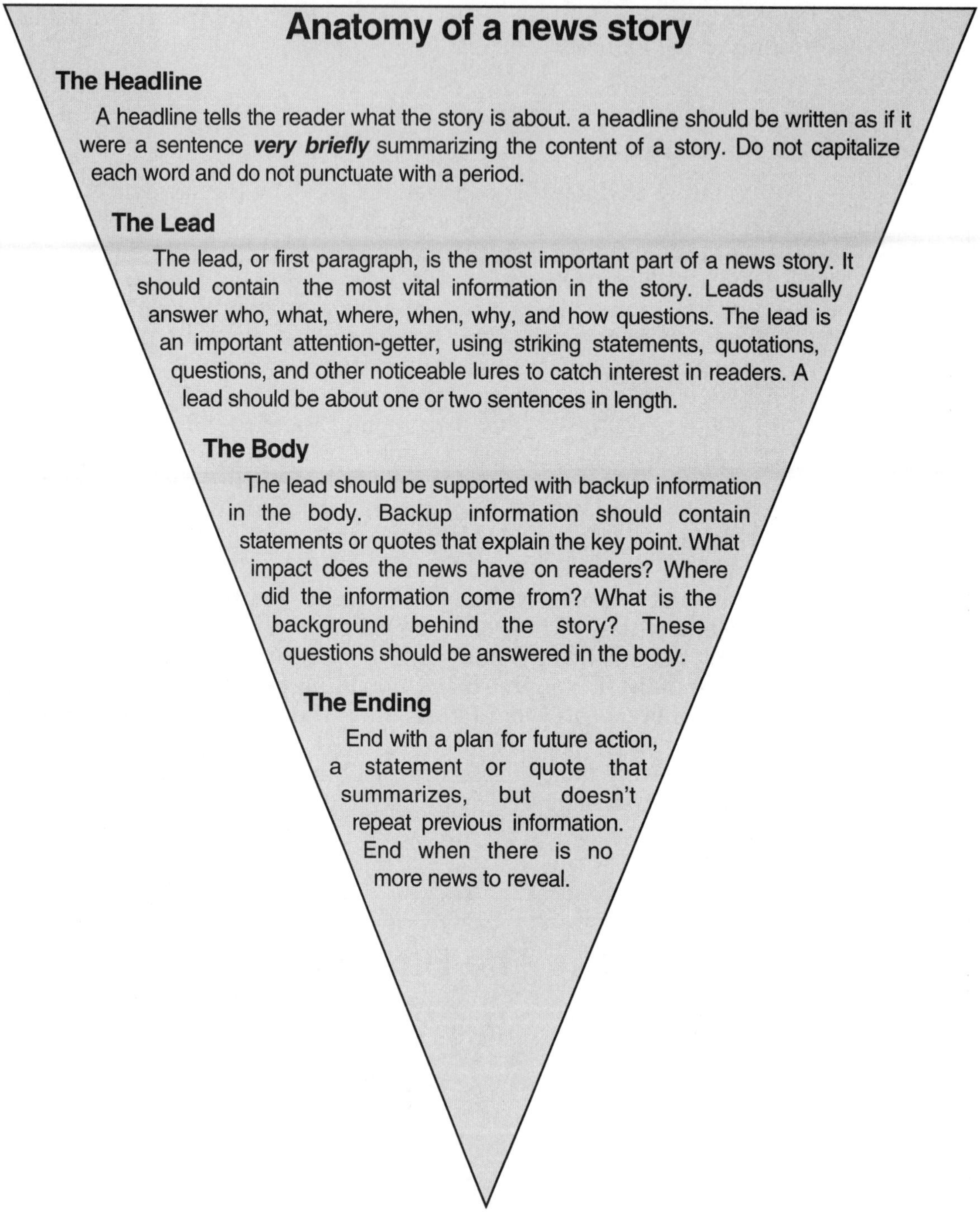

Anatomy of a news story

The Headline

A headline tells the reader what the story is about. a headline should be written as if it were a sentence *very briefly* summarizing the content of a story. Do not capitalize each word and do not punctuate with a period.

The Lead

The lead, or first paragraph, is the most important part of a news story. It should contain the most vital information in the story. Leads usually answer who, what, where, when, why, and how questions. The lead is an important attention-getter, using striking statements, quotations, questions, and other noticeable lures to catch interest in readers. A lead should be about one or two sentences in length.

The Body

The lead should be supported with backup information in the body. Backup information should contain statements or quotes that explain the key point. What impact does the news have on readers? Where did the information come from? What is the background behind the story? These questions should be answered in the body.

The Ending

End with a plan for future action, a statement or quote that summarizes, but doesn't repeat previous information. End when there is no more news to reveal.

FIGURE 3-8   Anatomy of a News Story.

8/8/02

Contacts:
Colin Johnson, Iowa Pork Industry Center
Sherry Hoyer, Iowa Pork Industry Center

Pork Producers Invited to Pork Quality Assurance Meetings

AMES, Iowa—Pork producers are invited to become certified or renew their certification in the Pork Quality Assurance (PQA) program at one of a series of meetings to be held across Iowa. Program sponsors are Iowa Pork Industry Center (IPIC,) Iowa Pork Producers Association and Iowa State University (ISU) Extension. Eighteen meetings are scheduled from Aug. 26 to Sept. 12. Dates, times and locations for each meeting are listed below.

IPIC extension program specialist Colin Johnson said the meetings will provide opportunities for all pork producers to learn more about the PQA program, animal handling biosecurity, herd health, food safety and other related topics.

"These programs are for all pork producers, whether they're becoming PQA certified for the first time or are recertifying," Johnson said. "In answer to consumer demand, packers expect their producer clients to be certified and these programs will help in that effort."

Speakers will vary by site, but several programs will include Johnson and IPIC associate director James McKean. McKean also is ISU Extension swine veterinarian. ISU Extension livestock and swine field specialists will present information at programs in their respective areas. Veterinarians also are encouraged to attend with their respective producer clientele. Although the educational content of the PQA program can be achieved during these meetings, veterinarians need to sign the official certification document for each of their producers.

There is no cost for producers to attend the meetings or to obtain PQA certification. For more information, contact your local county ISU Extension office.

(continues)

Meetings are scheduled as follows:

Aug. 26, 1 p.m., Villisca Senior Center

Aug. 26, 7 p.m., Harlan, Shelby County Hospital, Aubel Room

Aug. 27, 7 p.m., Cylinder Community Center

Aug. 28, 10 a.m., Correctionville Community Center

Aug. 28, 1:30 p.m., Cherokee, Western Iowa Tech Community College

Aug. 28, 7 p.m., Storm Lake, Buena Vista County Extension Office

Aug. 29, 1 p.m., Toledo, Tama County Extension Office

Aug. 29, 7 p.m., Iowa City, Johnson County Fairgrounds, Montgomery Hall

Aug. 30, 1 p.m., Ottumwa, Southern Prairie AEA, Hwy. 63 N

Sept. 3, 1 p.m., Burlington, Des Moines County Extension Office

Sept. 3, 7 p.m., Washington, Pizza Hut

Sept. 4, 7 p.m., Ames, ISU College of Veterinary Medicine, Alumni Room

Sept. 5, 1 p.m., Sac City, Iowa State Bank

Sept. 5, 7 p.m., Carroll Recreation Center

Sept. 9, 1:30 p.m., Peosta, Northeast Iowa Community College, National Ag Safety Center

Sept. 9, 7 p.m., Winthrop, St. Patrick's Church

Sept. 12, 1 p.m., Waterloo, Hawkeye Community College, Tama Hall

Sept. 12, 7 p.m., Mason City, North Iowa Fairgrounds, 4-H Learning Center

(Permission granted and adapted from Iowa State University Extension.)

This example starts with standard formatting conventions.

8/8/02   The release date of the news.

Contact: Colin Johnson, Sherry Hoyer—Iowa Pork Industry Center   the contact person, who can clarify any information presented in the release to a newspaper or magazine. The contact person could be the writer or someone closely tied to the news contained in the release. The key is to put the name of the person who can articulate effective answers for any issues that may arise.

Pork Producers Invited   a concise title designed to communicate a message and draw in the reader to explore the content of the article.

First paragraph   a good example of a strong lead with all of the pertinent information.

Who?   Iowa Pork Industry Center, Iowa Pork Producers Association, and Iowa State University

What?   Certification in the Pork Quality Assurance program

Where?   Assorted sites across Iowa

When?   August 26–September 12

Why?   To provide information to Iowa pork producers about certification and recertification information for the Pork Quality Assurance program

How?   Through a collection of meetings across Iowa, pork producers across Iowa can learn about the quality assurance program.

Rest of the news release   a collection of facts and quotations assembled to support the lead paragraph. It tells the reader more about the basic information presented in the lead paragraph. It also tells pork producers how to get more information

Schedule   a complete schedule of the quality assurance meetings.

This release was written in inverted pyramid style. As the style dictates, secondary facts were reserved for the end of the article. If a newspaper editor deems it too long for the space allocated, the last paragraph (or even more) could be deleted without affecting the tone of the article.

THE PARTS OF NEWS

Aside from the lead, a news article has a number of other components that help readers understand what is being communicated.

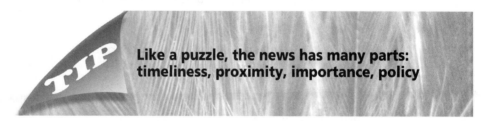

Like a puzzle, the news has many parts: timeliness, proximity, importance, policy

The first set of components is called **news determinants**. Determinants are a set of criteria that writers and editors use to decide if news is fit to write and print. The determinants are, in no particular order, timeliness, proximity, importance, and policy. In order for a story to be printed, it must meet the fundamental test that each of these determinants poses.

Timeliness   the freshness of the news story. A piece about the advent of tractors using diesel, for example, would be out of date. However, if a local seed company just introduced a pest-resistant variety of corn seed using genetic engineering, writing an article on it would be perfect timing.

Accurate news stories must have an element of timeliness.

Proximity   the term that describes where a news story took place and the people it affects. Presidential speeches, congressional action, and monumental emergencies, for example, are national in proximity. Other events are relevant only to a particular locality. A report on the yearly cotton crop in South Texas, for example, would not be of much interest to newspaper readers in southern Illinois. Most likely, they want to know about the crops in their area.

Importance   is every news story or subject is important to someone. But magazines, radio broadcasts, Web sites, and newspapers have only a finite amount of space at any given time, so editors must choose carefully what to publish. That is where the concept of *importance* as a news determinant applies. Because space is limited, editors must pick stories that are of importance to the most people in their audience.

Policy   Most publications like magazines and Web sites are focused in the subjects they pursue; their policy is to reject any stories that do not fit their editorial goals. For example, a piece about a standout high school athlete might work well in *Sports Illustrated*, but will it be valuable to the readers of *Progressive Farmer*? Briefly, a publication's editorial policy is the rulebook that editors follow and is ultimately the final determinant in whether a piece is used.

In addition to news determinants, the subject of an article is governed by **news components**. News components are conflict, unusualness, progress, and human interest.

All news stories or feature stories are based on the four components listed. In the news release on pages 55–56, the news component was based on progress: it highlighted Iowa's Pork Quality Assurance program. A feature article that highlights a historical figure and his contribution to preserving the history of cattle ranching in Texas is based around a human-interest aspect.

Following is a description of the components:

Conflict   Conflict stories revolve around the struggles faced by people everywhere. In the mainstream press, stories of conflict may be about Congress's debating a bill. Conflict may also be a community's fight to make its water safe by boycotting a local industry. In agriculture, conflicts also exist. A group of local farmers may be dealing with bad weather or flat markets. How the cattle industry faced the recent threat of mad cow disease is another example.

Unusualness   Anything out of the ordinary qualifies under the unusualness heading. In agriculture, unusualness may be described as the DNA cloning of a sheep or a farmer who grew a twenty-pound tomato.

Progress   Progress as a news component is a natural as a news story. Everyone likes to hear stories of progress, as it affirms our belief in the strength of the human race. Progress may be a football team's meteoric rise to success or the local factory's hiring 100 new employees. In agriculture, progress is made almost daily (Figure 3-9). Seed companies are constantly releasing new, higher-yielding seed varieties, and the genetics of livestock improvement are improving at an exponential level. From a news standpoint, agricultural progress is what most stories in magazines are about. They give insights—for example, how a producer cut costs and began a grazing program that led to higher weaning weights on his calves.

Human interest   stories that connect emotionally with readers. People like to read stories about other people to whom they can relate. Many times, folks want to read about ordinary people like themselves and the lives that others lead. Examples of human interest stories might be a farmer who is also an accomplished classical musician or a woman rancher who is at the top of a field typically dominated by men (Figure 3-10).

FIGURE 3-9   Agricultural progress is made almost daily, as in the case of livestock genetics. (*Courtesy Russell A. Graves*)

FIGURE 3-10   A working cowboy and his daily chores is an example of a human interest story. (*Courtesy Russell A. Graves*)

News components and determinants are the parts of a story that make it relevant to both readers and those publishing the piece. A good litmus test when thinking of writing an article or news story is to determine if it meets the criteria as determined by a news component or determinant.

WRITING THE FEATURE STORY

Like a news story, the feature story must start with a strong lead in order to connect with the reader. Unlike the news story, however, it does not have to be as concise. In fact, most magazine articles are in the 1,000- to 3,000-word range.

Feature stories allow writers to expand on a subject in much more detail than is allowed in news stories. Feature articles are generally used in magazines and special sections of newspapers and report on human interest, conflict, progress, or unusualness.

Publications that use feature stories often look for a wide variety of subjects since they seek to appeal to as broad a readership as possible. For example, a quarter horse magazine might want to publish feature stories ranging from breeder and horse profiles to where-to-go-riding pieces that give readers hints on how to get the most out of their horse (Figure 3-11). The magazine may not want feature stories on cotton management strategies in the Deep South, although that topic would certainly be appropriate for a grower magazine based in that region of the country.

The right way to go about seeing what kind of feature stories that a particular publication likes to use is to do research. Get several back issues, and read the articles to get a sense of the publication's style as well as the types of stories it publishes. Then you can make an educated guess as to if you have the ability to write for the publication and the types of pieces it might like to use.

ELEMENTS OF A GOOD NEWS STORY

(Permission granted to reprint by the Texas Tech University Department of Agriculture Communications.)

1. Cover most, if not all, of the five W's and H.
   - WHO will do, said or did something
   - WHAT will be done, was said or happened
   - WHEN it will be done, it was said or happened
   - WHERE it will be done, it was said or happened
   - WHY it will be done, it was said or happened
   - HOW it will affect me or how it was done
2. Follow the inverted pyramid style with most important facts first.
3. Keep your sentences short.
4. Use short, well-known words (avoid jargon).
5. Assume no one knows anything about your subject, but don't insult anyone's intelligence.

FIGURE 3-11 Because magazines often try to appeal to the broadest readership possible, a horse magazine may include features, profiles, and hints. (*Courtesy Russell A. Graves*)

6. Use active words to add zest (avoid, for example, "I said," or "I thought").
7. Use specific, concrete—not abstract—words and terms.
8. Do not editorialize (inject your own views or those of your source without attribution).
9. Avoid color adjectives, those that reflect opinion.
10. Keep paragraphs short. Usually two or three sentences are enough, and sometimes you may have only one sentence in a paragraph.

11. Create easy-to-read paragraphs that amplify the lead or preceding paragraphs.
12. Write in third person (he, she or John said). Second person is becoming more common.
13. Know when to quit writing.
14. Avoid pronouns for which the antecedent is unclear (use nouns to avoid ambiguity).
15. Avoid introducing sentences with dates or prepositional phrases (and avoid overuse of prepositions inside the sentence).
16. Proofread copy and edit unnecessary words; eliminate errors of grammar and spelling.
17. For good measure, have someone else read the copy and tell you what they think it said.
18. If news story is longer than one page, write "more" at the bottom of each page.
19. Indicate the end of the news story by centering the characters —30— or ### below the end of a story.

To Keep Your Credibility

1. Correct grammar, punctuation, spelling (especially names).
2. Check facts such as titles, dates and figures twice—when you are writing and afterward.
3. Have someone else read your story to check for typing errors and comprehension.
4. Rewrite awkward sentences—eliminate extra words.
5. Re-read the story one last time to see if thought and fact follow logically.

THE FEATURE STORY: GETTING STARTED

In Chapter 2, we discussed how to conduct sound research so you can get all of your facts straight when working on a written piece. One way to make your writing more effective and less troublesome is to use an outline.

You probably learned at some point in school that an outline is an excellent way to organize thoughts when writing. It is still true. Using an outline, especially when working on long pieces such as feature articles, is an extremely effective way to keep your writing flowing from one point to the next without being redundant.

An outline begins with a main topic. Once a main topic is established, next come the main points, followed by the subpoints. The subpoints, if need be, can be further supplemented with minor points to clarify an issue brought up in a piece.

A sample outline for an article follows.

Title: *The Life that Wouldn't Wait*

Main Topic: *Erwin Smith and His Contributions to Art and History*

I. The role of Erwin Smith and his importance to western culture
    A. Quote from a noted author
    B. The controversy surrounding Smith
II. The Life History of Erwin Smith
    A. His early years
       1. Personal family history
    B. His teens
       1. Travels to West Texas
       2. His love of the land
    C. His young adult years
       1. Travels back East
III. Smith enters art school
    A. Travels to Chicago
    B. Travels to Boston
IV. Smith travels back to the range
    A. His friendship with George Patullo
    B. The reason for his photographs
V. His later years
    A. His legacy of work
    B. His death
VI. Conclusion

As you can see, an outline organizes a writer's research and thoughts. To a lesser extent, it also serves as a psychological booster. Instead of facing the task of writing a 3,500-word piece from scratch, this writer can look at the outline and know he or she only has to write six 600-word components that, put together, make up the whole story.

EDITING YOUR WORK

As hard as it is to do, perhaps the most important aspect in writing is editing. Writers must get in the habit of rereading or getting someone else to reread their work.

The more you read a piece that you have written, the more likely you are to find ways to make it more concise and easy to read. Many writers

---

Deciding on what's relevent

As great as the Internet is as a tool for finding information, it does have some major drawbacks. The fact is anyone can put anything on the web whether it is true or not.

FIGURE 3-12   By reading and rereading a story, a writer can often find ways to improve the piece.

finish a piece and set it aside for several days so that the material becomes a bit stale in their mind. By waiting, they are able to detach themselves from the piece and then read it more subjectively.

When editing your work, look for the following elements.

- **Redundant phrases.** Don't repeat yourself over and over. Say it once, and say it clearly.
- **Clichés.** Phrases like "Time will tell . . ." are overused and should be avoided.
- **Colloquialisms.** Local jargon like "ya'll" and "fixin' " may be confusing to readers who are not local. Think nationally, and use words that people everywhere understand.
- **Unnecessary phrases.** Sentences that start with "There are" or "There is" are weak and should be revised.
- **Passive sentences.** "The fields were plowed by the migrant workers" works much better as an active sentence: "The migrant workers plowed the fields." Passive sentences often add unnecessary words.
- **Inaccurate facts.** Although you may have researched correct facts, you may have transcribed them wrong. Double-check for accuracy.

Editing is just as much an editorial process as writing is. In editing, writers must check their work for accuracy, hone a piece down to the required number of words, and rewrite within a set deadline (Figure 3-13).

Editing takes practice. A good place to start is by learning all of the copyediting marks and perusing newspapers for mistakes. Due to the speed at which newspapers are released, mistakes do happen.

CONCLUSION

The ability to write expressive and persuasive copy is a gift. Nevertheless, writers can learn the skills. Effective writing starts with the ability to rec-

FIGURE 3-13   Like writing, editing takes practice. (*Courtesy Russell A. Graves*)

ognize the importance of spelling and grammar. English and language classes should have taught you those lessons. This textbook helps you put those lessons into action.

Writing well and writing for pay can go hand and hand. The country is full of markets for good writers who can recognize the newsworthiness of a good story and report on its relevance to an audience. As you become proficient as a writer, you will put into action all the fundamentals that you learned in schooling. These fundamentals are called mechanics.

Once you have a good knowledge of the mechanics of putting a story together, you should learn professional writing techniques, such as using the inverted pyramid and crafting a strong lead to add professionalism to your work.

One of the last professional writing lessons you can learn is to edit your work and rewrite often. Rarely is a story perfect on the first try. Only after editing and rewriting will it become suitable for publication.

Writing is like playing a sport. You may be naturally talented, but to rise to the top of your game, you have to practice—and practice often. Only through structured and correct repetition will your work remain consistent.

REVIEW QUESTIONS

Multiple Choice

1. As great as computer software is, it should not be used as (for)

   a. a doorstop.
   b. the only means of editing misspelled words.
   c. gaming only.
   d. none of the above

2. Correctly spelled words can

   a. make people think you are smart.
   b. help you get a better job.
   c. yield power and efficacy to your communications.
   d. make it difficult to write an article quickly and efficiently.

3. Most articles in just about any popular magazine are written on a(n)

   a. college level.
   b. elementary level.
   c. high school level.
   d. junior high level.

4. The ABCs of journalism are

   a. accuracy, bravado, and clearance.
   b. accuracy, brevity, and clarion.
   c. acuteness, bravery, and clarity.
   d. accuracy, brevity, and clarity.

5. Brevity relates to the _____ of the article.

   a. length
   b. shortness
   c. conciseness
   d. precision

6. In determining clarity, the writer must ask himself or herself,

   a. "Why am I doing this?"
   b. "What is the intent of this article?"
   c. "Does what I just wrote make sense?"
   d. "Why is the sky blue?"

7. Clarity demands that the writer stick to the facts and

   a. editorialize as much as possible.
   b. state his or her opinion.

     c. not editorialize.

     d. give them in bulleted format.

8. Effective news releases start with a

     a. timid weak lead that will allow the reader to slowly work into the material.

     b. strong lead.

     c. lead that states the main subject of the release.

     d. none of the above.

9. The first paragraph of the piece presents

     a. none of the pertinent facts.

     b. some of the pertinent facts, with the rest given later.

     c. all of the pertinent facts.

     d. most of the pertinent facts.

10. Subsequent paragraphs of a news story provide _____ that relates to the lead paragraph.

     a. irrelevant information

     b. supporting information

     c. superfluous information

     d. all of the above

11. Using the inverted pyramid writing style ensures a

     a. confused reader.

     b. somewhat continuous flow of facts.

     c. continuity of facts and a naturally flowing means of providing information.

     d. bullet statement list of facts and a flowchart leading to the main point of the article.

12. Determinants are a set of criteria that writers and editors use to decide if

     a. news is fit to write and print.

     b. an idea is good enough to write about.

     c. the correct format was used in the writing of an article.

     d. the pyramid style should be slanted or inverted.

13. Timeliness refers to the

     a. speed with which the article was turned in to the editors.

     b. century in which the event happened that the article is written about.

     c. freshness of the news story.

     d. none of the above

14. *Proximity* is the term that describes where the

    a. writer lives in respect to where the news event took place.
    b. news story took place and whom it affects.
    c. writer was when the news event took place.
    d. all of the above

15. Since space is limited, editors must pick those stories that are

    a. the shortest.
    b. the shortest but are of least importance to the most people.
    c. of importance to the most people.
    d. the shortest regardless of how important they are.

16. Conflict stories

    a. revolve around the ease with which some people live as compared to others.
    b. revolve around the age-old struggle between good and evil.
    c. revolve around the struggles that people everywhere face.
    d. do not revolve around anything.

17. Human interest stories

    a. deal with only humans.
    b. are only about things that are interesting to humans.
    c. do not deal with humans.
    d. connect emotionally with the readers.

18. Feature stories allow the writer to

    a. expand on a subject in detail.
    b. feature his or her own ideas on the subject.
    c. write about only one subject.
    d. none of the above

19. Proofreading eliminates

    a. errors of grammar and spelling.
    b. the need for editors.
    c. the possibility of the story's not getting published.
    d. your responsibility of factual writing.

20. An outline is an excellent way to

    a. keep from having to write long articles.
    b. organize your thoughts when writing.
    c. express your opinions.
    d. keep your words in an organized arrangement.

Fill in the Blank

accuracy	quote	responsibility
descriptive	releases	written
detach	several	main
determinant	solid	subpoints
editorial	style	fundamental
events	unusualness	audience
feature	vary	importance
good lead	venues	sense
guess	wasting	timeliness
human	wordiness	determinants
human	begins	progress
industry	question	lead
longer	subjective	breaking
outline	reject	material
points	conflict	rule
policy	generates	types
powerful	intellectually	progress
proximity	inverted	appealing
publication	ability	opinion
pyramid	components	wasting
qualifies	interest	abbreviations

1. As an agricultural communicator, you have the _____ to represent your employer and the _____ in the best light possible.

2. Effective communicators should learn to express their thoughts fully without excessive _____, _____, or other nonconventional means of communicating.

3. When teamed with correct spelling, grammar is a _____ tool.

4. When preparing to write, you should always do an _____ assessment.

5. Since everyone's reading skills _____, magazines know that by _____ to a lower reading level, they can get more buyers and subscribers.

6. _____ is necessary in any credible communication whether it is written or spoken.

7. Accurate writing _____ with good, _____ research.

8. Web sites are unique in the fact that brevity must _____.

9. If you can catch a web browser's attention without _____ his or her time, then you can direct that person to areas that contain _____ content.

10. Clarity refers to the _____ word's ability to get the point of the article across to the reader.

11. Although editorial writing is appropriate for some _____, most news stories should avoid _____ and stick to the facts.

12. News _____ are tools used by corporations, small businesses, educational institutions, and other organizations to get the word out on _____ news and _____ that directly affect them.

13. A strong lead is what _____ attention for the story.

14. A _____ starts with the basics: who, what, where, when, why, and how.

15. The _____ _____ style of writing puts the most important facts of a story at the beginning of an article.

16. The news determinants are _____, _____, _____, and _____.

17. In order for a story to be printed, it must meet the _____ test that each news _____ poses.

18. Most publications are focused on the subjects they pursue; it is their policy to _____ any stories that do not fit their _____ goals.

19. News components are _____, _____, _____, and _____.

20. Anything out of the ordinary _____ under the unusualness heading.

21. _____ affirms our belief in the strength of the _____ race.

22. News _____ and _____ are the parts of a story that make it relevant to not only the readers but also the people publishing the piece.

23. A _____ story must start with a strong lead in order to connect _____ with the reader.

24. The right way to go about seeing what kind of feature stories that a particular _____ likes to use is to do a little research. Get several back issues and read the articles to get a _____ for the publication's _____ as well as the _____ of stories they buy. Then you can make an educated guess as to if you have the _____ to write for the publication, and the types of pieces they would like to use.

25. A(n) _____ is an extremely effective way to keep your writing flowing from one point to the next without being redundant.

26. An outline starts with a _____ topic; next come the main _____ followed by the _____.

27. You may wish to finish a piece and then read it again _____ days later so that the _____ becomes a little stale in your mind. By waiting, you are able to _____ yourself from a piece for a bit and then read it with eyes that are more _____.

ENHANCEMENT ACTIVITIES

1. Write at least five lead paragraphs on local news stories using who, what, where, when, why, and how in each paragraph.

2. Using the inverted pyramid style, pick a lead paragraph in Activity 1, and complete a news release.

3. Collect news releases from the Internet, and break them down into the elemental parts of who, what, where, when, why, and how.

4. Create a portfolio of sixteen newspaper and magazine articles that are based on the news determinants listed in the chapter.

5. Using the guideline listed in the section "Elements of a Good News Story," pick a topic, and write an outline and a feature story at least 1,500 words in length.

6. Once you have completed Activity 5, exchange your article with a classmate and edit each other's work, taking close note of spelling, grammar, and style.

WEB LINKS

To find out more about the topics discussed in this chapter, visit these Web sites.

hodu.com YOUR COMMUNICATION PORTAL
<http://www.hodu.com/>

Dictionary.com
<http://www.dictionary.com>

Thesaurus.com
<http://www.thesaurus.com>

Guidebook—Agricultural Communications in the Classroom
<http://www.oznet.ksu.edu/>
  (Search terms: rutherford, guidebook)

Preparing News Releases
<http://www.publicdoman.com/>
  (Search term: rotary PR)

12 Point Checklist for Writing Feature Articles
<http://www.netwrite-publish.com/>

Writers' Toolkit
<http://www.authorpower.com/>

Alternatively, do your own search at <http://www.alltheweb.com>
  (Search terms: effective communications, effective writing, journalism
    basics, preparing news releases, feature story, editing)

BIBLIOGRAPHY

Burnet, Claron, and Tucker, Mark. (1994). *Writing for Agriculture: A New
   Approach Using Tested Ideas* (2nd ed.). Dubuque, Iowa: Kendall/Hunt.
Hartenstein, Shannon. (2002). *Preparing for a Future in the Agriculture Com-
   munications Industry*. Manhattan: Kansas State University.
Instructional Materials Service. (2002). *Agriculture Communications 315
   Course Materials*. College Station, TX: Author.
National FFA Organization. (2003). *National FFA Agriculture Communica-
   tions Career Development Event*. Indianapolis: Author.
Sebranek, Patrick, Meyer, Verne, and Kemper, Dave. (1990). *Writers INC, A
   guide to writing, thinking, and learning* (3rd ed.). Burlington: Write Source.
Texas Tech University. (2003). *Ag Communications CDE Preparation Guide-
   book*. Lubbock: Author.

4

THE ELECTRONIC MEDIA

OBJECTIVES

After reading this chapter, you should be able to:

- understand the basic history of computer use and the Internet.
- identify the importance of the Internet for business use.
- search effectively for Internet resources.
- use e-mail and the Internet as a way of communicating.
- write effective e-mails.
- develop a Web page concept.
- understand the basics of Web design.
- identify successful selling tactics on the Internet.
- market Web sites using e-mail, the Internet, and other resources.
- understand the need for keeping Web sites current.

KEY TERMS

e-mail	IP address	target marketing
general marketing	link	World Wide Web
internet	server	

OVERVIEW

The computer revolution began during World War II. Scientists discovered that by sending electric currents through vacuum tubes, they could do complex mathematic computations with machines. The first computers, huge machines housed in large rooms, had about the same computing power as a digital watch.

Over the next few decades, the science of semiconductors was established and refined, and computers became smaller and smaller. In the early 1980s, the personal computer, or PC as it is called today, became commonplace. It brought computing power into the homes of people everywhere (Figure 4-1). Much smaller than the first computers developed, the early PCs still lacked large-scale computing power and required users to learn a computer language in order to perform operations.

In the 1980s, the Microsoft Corporation introduced the Windows operation system. Windows, as it is commonly known helped revolutionize the computer industry by making computing as easy as pointing a mouse at an object and clicking on it to make a program run.

In the 1990s, the computer became a household staple—a tool as important as other appliances in the home. Because of the ease of use and the increased computing power, millions of households acquired PCs for a variety of uses, including desktop publishing, home finance, and word processing.

FIGURE 4-1   The electronic revolution that created the PC brought computers into homes and businesses everywhere. (*Courtesy USDA/NRCS*)

Since the mid-1990s, society's use of communicating electronically has increased dramatically. Since 1990, the use of the **Internet** has increased 35,000 percent! In fact, by early 2002, half of all American households were connected to the vast resources that the Internet offers.

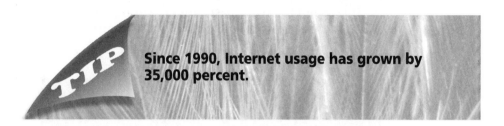

Since 1990, Internet usage has grown by 35,000 percent.

Like other industries, agriculture has become fully immersed in the electronic World Wide Web. A search of the Web reveals news sites, electronic commerce sites, universities, research stations, and a host of other agri-specific topics. Finding information about obscure breeds of animals or the latest genetic technology is as easy as logging on and pointing the Web browser to the appropriate sites.

Agriculture's use of the Internet is crucial for the industry to stay a relevant and viable industry in today's fast-paced business climate. Both agribusiness and production agriculturists must embrace the medium to stay competitive.

This chapter contains information on how to increase agriculture's presence on the Internet, thus increasing the productivity of ranchers, farmers, researchers, and business owners. Specifically, you will learn how to use the Internet for facilitating communication for a variety of reasons. In addition, this chapter walks you through the basics of Web site design and marketing the Web site once it is created. Finally, it explains why keeping Web sites fresh is important for the productivity of a business.

AN INTRODUCTION TO AGRICULTURE AND THE ELECTRONIC MEDIA

Since agriculture has always been on the cutting edge of technology, it makes sense that the nation's oldest as well as the newest industry are perfect partners. Agriculturists have a profound interest in keeping current on the latest Web technologies when it comes to information dissemination, market tracking, and electronic commerce (e-commerce) ventures. E-commerce is used to sell products and services over the Internet with no person-to-person contact required.

Successful producers and agribusiness recognize the power of the Internet and **e-mail** as a tool for increasing their productivity and profitability (Figure 4-2). Therefore, successful agricultural communicators must have the knowledge to use electronic communication in every possible way.

It is useful to have a basic understanding of how the Web works so that you can use it to its fullest extent as a powerful business tool. By knowing

FIGURE 4-2   Successful producers and agribusinesses recognize the power of the Internet. (*Courtesy USDA/NRCS*)

about the Internet, you can effectively recognize its potential as well as its limitations.

What Is the Internet?

The Internet is a vast network of computers that store information (Figure 4-3). Users can connect to the Internet and get information from any computer that allows free access (some Web sites have controlled access for business privacy, and others charge for access).

A highway analogy is appropriate for this discussion. Imagine that each site on the Internet is a destination that you could physically visit. In order to get to your destination, you have to get in your car, drive down the road, arrive at your destination, and go in the front door.

The Internet works the same way. In the cyberworld, the computer is the car, the worldwide network of phone lines serves as the road, and a company's Web site is the destination that your computer takes you to. You direct your computer where to go by typing in the appropriate address (e.g., <http://www.russellgraves.com>). Instead of looking for the words you type in to a web browser, the computer looks for an **IP address**, a series of numbers that identifies every computer connected to the Internet.

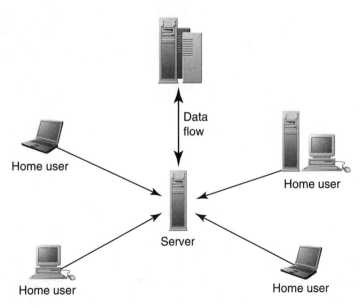

FIGURE 4-3   The Internet backbone.

Servers store the information that you want to access.

In essence, a computer is an access point for the Internet. **Servers** store the information that you want to access. Consider this: If all of the information on the **World Wide Web** was stored on one server and the server quit working, the entire Internet would quit working (that is the reason that Web sites are stored on thousands of different servers all over the world). But, if *your* computer quit working, the Internet, and everyone using it, would be unaffected.

How Does E-mail Work?

E-mail works much in the same way that Web sites work. To contrast the two, surfing the Web involves looking for information, while crafting e-mails is sending information.

A typical e-mail address may be something like <bob@anywebsite.com>.

A typical e-mail address may be something like: <bob@anywebsite .com>. In this example, *Bob* refers to the user's name. The @ symbol means simply "at." And *anywebsite.com* is the Web site that hosts Bob's mail account. The *.com* on the end of the site's name indicates that it is a commercial, or business, site.

Whenever an e-mail is sent to <bob@anywebsite.com>, the mail is first routed from the sender's server to the anywebsite.com server. Once it is received by anywebsite.com, the server computer looks for the name on the front of the address. If it is a large company, the mail server may have a thousand names entered into it. The mail server software, though, looks at the message, sees that it belongs to Bob, and puts it in Bob's folder for temporary storage.

The e-mail stays on the server at anywebsite.com until Bob logs on to his mail account. At that point, the mail is downloaded on to Bob's computer, where he can read the message, print it, save it to a file, delete it, or respond to the person who sent it.

E-mail stays on the server until it is downloaded, where the receiver can read the message, print it, save it to a file, delete it, or respond to the person who sent it.

Searching for Resources

As Web sites began to multiply exponentially, it became next to impossible for users to find what they were looking for without the aid of search engines. Search engines are an on-line resource for locating Internet resources.

Search engines work by letting users type keywords into a text box on screen (Figure 4-4). When the user clicks the Search key, the search engine looks through its database for possible matches to the keywords typed in. The search software then displays the results by prioritizing them into the most likely matches. The search engine software constantly updates the search server by continuously looking for new sites on the Internet.

For businesses, search engine ranking is one of the primary keys to their Web site's success. It is unlikely that people searching for a Web page about wheat will care to look beyond the first twenty sites listed by a search engine. Therefore, the higher the ranking is on a search engine, the more likely it is that traffic will be directed to that site. Like traditional brick and mortar businesses, traffic into the business is essential for keeping the business solvent and open.

Deciding What Is Relevant

As great as the Internet is as a tool for finding information, it has some major drawbacks. The fact is that anyone can put anything on the Web, whether it is true or not. The key to making the World Wide Web a valuable resource for finding information is not hard. Consider these points.

- **Stick to reputable Web sites for factual information.** Most agricultural magazines maintain a Web site. If you trust what appears in a publication, you probably can trust its Web site too.

FIGURE 4-4   Search engines help users wade through the mounds of information on the Internet in order to find specific information—for example, on cotton. (*Courtesy Russell A. Graves*)

- **Learn to discern fact from opinion.** Fact is fact and cannot be disputed. Opinions are interpretations of facts and can vary greatly from person to person. Learn to recognize which is which. Do not always assume that one's opinion is factually correct.

TIP

Fact v. Opinion: Which is Which?

- **Double- and triple-check all information against other Web sites as well as printed documentation.** Good agricultural communicators verify all information from other, nonrelated sources. (Think about it. What would take more time: fact-checking or working to undo factually incorrect information that the company you represent has already disseminated?)

- **Keep track of the good sites, and do not revisit the bad ones.** If a Web site has proven itself to be accurate, save it to your "favorites" list. Forget about those that are factually inadequate.

Using the Internet and E-mail for Communicating

Perhaps the most beneficial aspect of electronically connecting with computers worldwide is the speed and efficiency at which communicating with other individuals takes place: An important e-mail message typed in Illinois can reach its intended recipient in Japan in just a few seconds. Moreover, the e-mail has a trail that allows users to track when the message was sent, an e-mail receipt to see if the message has been read, and then any subsequent replies to the original e-mail. With a minimal amount of keystrokes, users have a complete record of all correspondence to a person on the other side of the world (Figure 4-5).

The Internet can also be a valuable tool for communicating. Suppose you are the communications director for an agribusiness enterprise and one of your key responsibilities is to disseminate news releases about the company. Using the Internet, you can post the releases on-line so the

FIGURE 4-5 The Internet allows quick and easy communication with the world. (*Courtesy Russell A. Graves*)

target audience can download the information when they need it, resulting in time and money savings for your company because you no longer are confined to traditional mailings through the postal service.

Effective use of e-mail can save a company money.

Attaching Files

An effective use of e-mail is attaching files to correspondence. Many magazines now accept article submissions as e-mail attachments, thus saving the writer postage and time when it comes to beating deadlines.

E-mail attachments come in many forms: text files, photographs, videos, presentations, and spreadsheet files. The first key to sending attachments is to understand the software that created the attachment.

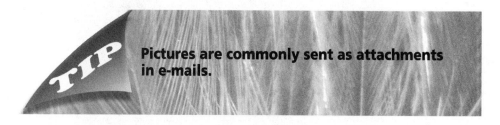

Pictures are commonly sent as attachments in e-mails.

If you created a text file on an obscure word processing program, can your recipient open the attachment? Communicate ahead of time, and find out the kinds of attachments that the recipient is able to use. Software and computer compatibility are essential in sending files attached to e-mail. An unopenable attachment is worthless to the recipient.

A word of caution: The most common way that computer viruses are spread is by opening attachments. A virus is a file that runs on a computer that causes problems for the computer owner on various levels. Mild viruses may simply e-mail everyone in a computer's address book a silly message. Others may destroy data, rendering the material unrecoverable.

Whenever you send or receive attachments, always scan them using antiviral software. Receiving tainted e-mails can ruin your computer and compromise your data. Sending them can cause hard feelings among your business contacts and cost you hard-earned business.

SELLING ON THE INTERNET

Selling on the Internet has made millionaires out of many people. Web entrepreneurs identify needs in the marketplace and occupy a niche. They fill a void in the marketplace, and consumers rush to buy the products or services. One such successful business is the on-line bookseller amazon.com. Amazon.com revolutionized the book selling business by offering a huge on-line inventory, complete with book excerpts and reviews as well as sales rankings. Customers log on to the Web site; search for a book by title, keyword, or author; browse for a selection; and order. From the comfort of their home, users of amazon.com can order any time of the day, and the books are shipped to the address of their choosing.

The advantage of selling on the Internet is that on-line stores never close: They can sell products all of the time. This aspect has changed the way people shop. Instead of going out for goods, it is conceivable that all you may need to survive can be purchased through various Web sites and shipped to your home (Figure 4-6).

E-commerce Web sites and the products they sell are limited only by a Web designer's imagination. If you can think of a product, it is probably

FIGURE 4-6   With on-line stores, you can buy anything—even hay! (*Courtesy Russell A. Graves*)

sold on the Internet. Agribusiness too has jumped on the cyber-band-wagon and embraced the Internet as a way of conducting business.

As a rule, e-commerce sites must be based around secure encryption so that a customer's private information, including credit card numbers, will be protected from criminal computer hackers.

Not only can an agribusiness use the Internet to sell products securely on-line, but also Web sites are key for selling a company to prospective customers. An agricultural service company can have a constant presence in cyberspace and outline every detail of the company to prospective customers.

THE SECRET OF EFFECTIVE E-MAIL

The secret of effective e-mail is really no secret at all. For most, the convenience of e-mail lies in its speed. Sending and receiving e-mail is quick, and you can get e-mail almost everywhere.

Effective e-mail gets its point across quickly and concisely. Since e-mail is hugely popular, many people receive dozens of e-mails during a workday. Wading through all of that electronic communication can take a serious bite out of a person's time. When writing professional electronic messages, keep that point in mind. To craft effective e-mails, here are some tips (Figure 4-7).

- **Keep it simple.** This isn't the place for colorful writing.
- **Get to the point.** If you can get your message across in a single sentence, do it. Don't waste someone's precious time by beating around the proverbial bush.
- **Check spelling and grammar.** This point is perhaps the most important. Sloppily spelled e-mails and poor grammar can tell a business

---

- Keep it simple.
- Get to the point.
- Check spelling and grammar.
- Always reply.

FIGURE 4-7   The Secret of Effective E-mail.

contact that you are less than professional and shake their confidence in your ability to complete a job.

- **Always reply.** If you get e-mail from a business contact, ALWAYS RESPOND. Even if there is apparently nothing in the message that needs responding to, send a reply to let the sender know that you received the message.

TIP

Effective e-mail is easy once you know the secret.

THE BASICS OF WEB PAGE DESIGN

There is no big secret for making Web sites profitable. In fact, it is Web site marketing and name recognition that drive the profitability of Web sites. Some Web sites are huge, with complicated navigational structures, while others are simple and take up only a few pages. The key to the success of Web sites of both kinds lies in the overall design (Figure 4-8). Successful Web sites are easy to navigate and user friendly; all of the text is easy to read and of high contrast; and the artwork is original, as well as sharp and clear.

For the agricultural communicator, the ability to decide on what content to put on a Web site and then deciding a concept around the information is as practical a skill to know as any other. In order to build a Web site, you do not necessarily have to know complex programming languages since many software packages are visually driven and write all of the code once the layout is designed. This is not to say that learning Web design languages is not important, because it is. However, a discussion on Web programming languages could fill up an entire book.

Deciding on Content

Deciding on content is perhaps one of the most challenging aspects of designing a Web site. The designer undoubtedly has more than enough information to promote, but Web space costs money, and it is up to the designer to decide what goes on the site and what does not. Deciding on content requires analyzing what it is that the business has to offer.

Suppose you are building a Web site about a beef cattle ranch. Here are some things that would be relevant content to add to the Web site:

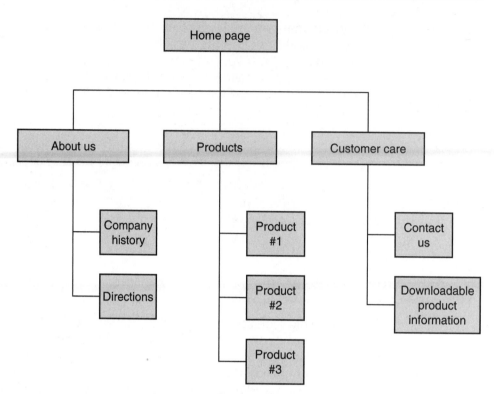

FIGURE 4-8   The navigation structure of a Web site should be easy to visualize and navigate for Web browsers.

- Ranch location
- About the ranch/history
- Breed of cattle raised
- Livestock available for sale
- Supporting information as to why this ranch's cattle would be worth buying
- Contact information

All of this information is relevant to the business. Each bit of information supports the main goal of the business: selling cattle.

Just as important as knowing what kind of information to add to a Web site is knowing what to leave out. The Web is full of sites with information that is useless to anyone but the Web site owner and his or her closest friends and family.

To give an example, suppose the owner of the beef cattle ranch has an outstanding collection of stuffed animals. The owner may be proud of this collection, but it has no place on a Web site about selling cattle. If the stuffed animals are for sale, then it may be appropriate to build another site just for the animals.

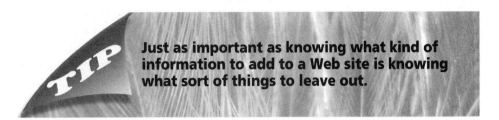

Just as important as knowing what kind of information to add to a Web site is knowing what sort of things to leave out.

If you are working for a client, you are relegated to putting on the Web site what the client wants, not what you think is important. The best way to approach this situation is to communicate with the client often as to what the content of the Web site will be. Part of the job of Web site designer is to guide the client to understand what is relevant for the Web site and what is not.

Designing a Concept

Once the content has been decided on, the next phase is designing the concept. It is hard to talk about design concepts because of the abstract nature of designs. Good designs are better seen than talked about anyway. Nevertheless, there are some basic design considerations when coming up with a successful Web concept.

- **Make the site's design match the business.** Old English lettering may be artistic looking—but it may not be appropriate for a Web site for a feed company. Fonts such as Arial, Times New Roman, or Century Gothic are fonts appropriate for a wide variety of sites because of their simple designs and high degree of readability.
- **Keep the backgrounds simple.** The Web sites of the biggest corporations all have a design component that is nearly universal: simple backgrounds. When designing a Web site, keep the background simple by using basic colors, such as black or white, as a background. Stay away from clip art or photographs. Their varying colors can make fonts hard to read.
- **Make navigation easy.** The navigational structure of the Web site should have **links** so that a user can access every page of the site from any other page. To save file size and make Web site maintenance more manageable, think about using multiple file folders in which to organize a site.

- **Use big, simple fonts that contrast with the background.** It is frustrating for web users to look at a site that uses hard-to-read fonts. When planning a Web site, stick to basic fonts such as Times New Roman, Verdana, or Arial. Also, be sure to make the type color one that contrasts with the Web site background. Fancy fonts may be appealing, but remember that one important aspect to designing a Web site is to make it accessible to as many people as possible. Suppose a prospective client from another country ran across your Web site and saw that it was written entirely in calligraphy. That person may not know how to read the letters and probably will go to a more accessible site.

Use easy-to-read fonts. Complicated fonts can be hard to read.

- **Use artwork—but don't overuse it.** Clip art and photographs spice up a Web site, but there can be too much of a good thing (Figure 4-9). Unless the page needs photos or clip art to illustrate a concept

FIGURE 4-9   Too much clip art can be a bad thing.

or highlight products for sale, limit the artwork to about two pieces per page. Gratuitous use of artwork makes a site look cluttered. Stay away from animated clip art. It may be cute but serves little purpose.

- **Think clean.** Ultimately, a good Web site is a professional representation of a chosen business. Suppose you own an animal health store. When customers enter the store, you want them to see a business that is clean, uncluttered, and easy to get around in. The same concept applies to Web design. Web sites should be clean, uncluttered, and easy to navigate. Anything less, and users might go somewhere else.

Putting the Web Site Together

Web authoring software ranges from easy to complex. For beginning Web authors, Microsoft Publisher, Adobe Go Live, and Microsoft Front Page are good choices because their interface is graphical and allows you to place images and text where you want them by simply pointing and clicking. These simple interfaces do not work well for complex Web sites involving lots of animation or secure areas. Front Page and Go Live allow designers to create code in addition to its graphical interface.

Publishing to the Internet

Once the Web site is complete, the last step is to publish the work to the Internet. A word of caution: Always make sure the Web site is error free before publicizing it.

Web authors most often use FTP as a means of transferring data from their computer to the Web server. FTP (which stands for *file transfer protocol*) is used for moving files from one computer to another.

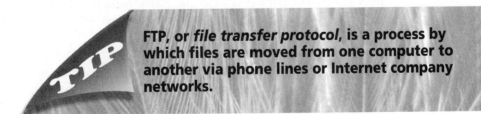

TIP

FTP, or *file transfer protocol*, is a process by which files are moved from one computer to another via phone lines or Internet company networks.

When files are uploaded via FTP, the person transferring the files must have permission from the Web server in order to do so. This permission comes in the form of a user name and password. Technicians create space on their server for the Web site and set up security around the space by providing only this user access via a password.

If an FTP site is not password protected, then anyone can erase the Web site and put another up in its place. The time it takes to update Web sites via FTP depends on the connection speed of the computer and the size of the files being transferred.

MARKETING A WEB SITE

The most crucial part of Web site design is the marketing process. You could have the greatest and most functional Web site ever built, but if no one knows about it, it is not doing the company you represent any good.

This section provides ideas on marketing a Web site. Remember that the ways listed are not the only ways to market. Use your creativity, and see what else you can come up with.

The Need for Marketing

The need for marketing should be transparent. The movie *Field of Dreams* has the famous line, *"If you build it, they will come."* That premise worked for the Kevin Costner character in the movie, but it is not enough for entrepreneurs and agricultural communicators. Once a Web site is built, the real work begins as you and your company endeavor to let everyone know about it. Marketing is no easy feat, especially with the proliferation of Web sites.

Marketing a Web site can be broken into two distinct segments. The first segment is **general marketing**: the process by which nonspecific groups are targeted. For example, an agribusiness may put a billboard in the town where it is located and advertise the Web site. This general type of marketing targets everyone who looks at the sign. Perhaps it will cause people who do not necessarily have anything to do with agriculture to stop by the company office: They may want to see what the company does.

TIP Marketing involves targeting your audience and aggressively seeking out those people who have a genuine interest in what you are selling.

The other segment of marketing, **target marketing**, involves aggressively seeking out people who are potential clients and telling them of your Web address. For example, a Web site for a livestock auction can help ranchers preview cattle before auction day. In this case, target marketing would mean sending a letter or postcard to all of the ranchers in the area telling them of this new service.

Whether your approach to marketing a Web site is general or targeted, the key is to let everyone you possibly can know about the site. Companies often put thousands of dollars and hours of time into their Web sites, so the site must produce revenue in order to be successful. And remember that millions of dollars have been made and lost based simply on the success or failure of a web site.

Using E-mail to Market the Site

E-mail is an effective way to market a Web site in both a general and targeted way. A general method of e-mail marketing is by adding a link to the site at the end of your e-mail messages—for example:

To: Bob@anywebsite.com

From: Russell A. Graves

Date: 3/11/03

Subject: Our meeting

Dear Bob:

Just a note to let you know I appreciate your taking the time to meet with me the other day concerning my proposal. I hope we can work together soon.

Sincerely,

Russell Graves

Check out our newly designed Web site at www.russellgraves.com!

In this example, the marketing ploy is quiet. If Bob wants to look at my Web site, he can do so quickly by clicking on the link. The link is presented in a nonconfrontational manner—sort of in a take-it-or-leave-it manner.

This type of e-mail marketing can be effective if you include your link on every e-mail you send out. The reason is that the person getting the message may not need your goods or services but may know somebody who does. The goal is to get the news of the Web site out in front of everyone.

Another type of e-mail marketing is sending an e-mail message to a targeted list of prospective clients announcing the Web site. Many companies sell their name lists to other companies for use in marketing ventures. These unsolicited e-mails are commonly called SPAM, and sometimes they annoy more than they benefit. Ultimately, the choice is up to you and how you want to use e-mail in your marketing plan.

Using Web Sites to Market a Site

A beneficial way to market Web sites is by using other sites to draw attention to your own. There are several ways to market a Web site on the back of other sites.

The first way to market using Web sites is by the use of *banner advertisements*. Banner ads are usually paid advertisements that link directly back to your Web site. The ads are often placed at the top or bottom of a page and are usually made up of images and text that are meant to entice people to click on the ad and be taken to the site. Banner ads are usually strategically placed where the audience they reach is of the targeted variety. A good place to advertise a business selling plow parts might be a farming Web site.

In *reciprocal links*, someone puts your link on their Web site and you do the same. Since each owner of a Web site agrees to share links, there is usually no cost for this type of marketing.

Web rings are related Web sites that join to try to increase traffic for everyone within the ring. It works like this: Somewhere on your Web site, you put a link that points to the web ring's home. If someone clicks on that link, it takes him or her to a site that has links to all of the members of the web ring. The idea is that someone interested in the subject of the ring would explore all of the sites listed.

In Internet marketing, there is nothing that is guaranteed to be wildly effective. Sometimes marketing schemes work, and sometimes they do not. It is important to try all of the different avenues and, most important, keep trying.

Other Marketing Means

Aside from e-mail and the Internet, there are countless other ways to market Web sites. Large corporations understand that their Web site is as much a part of their corporate identity as their logo, so they constantly work to put their Web site in front of the public.

Agricultural communicators have a duty to try to promote their Web sites as much as they can. To that end, there are several ways to promote a Web site name.

- T-shirts
- Note pads
- Mouse pads
- Pens
- Screen savers
- Direct mailings

These items are promotional and are usually given away. Aside from promotional items, always put your Web site on all letterhead, business cards, Rolodex cards, magazine and newspaper ads, billboards, and the like.

THE NEED FOR FRESH WEB CONTENT

Once you have built a Web site and marketed it, your job still is not finished. In order to keep a Web site effective, it has to stay current. Staying current may mean having a constantly updating time and weather ticker on it. Alternatively, staying current may mean having stock and commodity prices listed prominently on the site.

Whatever the Web site content centers around must be fresh (Figure 4-10). People who need a Web site for information expect the latest a company has to offer each time they log on. Internet users who do not find fresh content will go elsewhere for the information. An exodus of computer users on your site can spell doom for the site and the business.

A good way to let users know about Web site updates is to build a database of names and e-mail addresses. The collection of names can be accomplished in two ways. One way is to set up a response box on the Web site where people can sign up for announcements that the site has been updated. Another way is to harvest the names of people who send e-mail to the site. By collecting names and e-mail addresses, you ensure a huge list of contacts to use in targeted e-mail marketing strategies.

CONCLUSION

There is no doubt: The Internet and e-mail have changed the way that Americans do business forever. Local markets have turned into state, national, and international markets as well as created a new class of home business entrepreneurs. No longer is it necessary for all professionals to commute to a traditional physical business in a car. Now, people can telecommute to a virtual store on a server halfway across the country.

FIGURE 4-10   Like fruit, Web sites need to stay fresh to remain useful. (*Courtesy USDA/ARS*)

With that in mind, agriculturists and agribusiness are in a key position for taking advantage of the vast wealth of information and e-commerce possibilities that the World Wide Web offers. To take advantage of all of the cyber-opportunities, agricultural communicators need to understand how to maximize the resources in the business they represent and have a sound understanding of how the Internet and e-mail works. Those who know how cyberspace works can design effective and aesthetically sound Web pages that reach millions of people.

Once Web sites are built, marketing is key. To market Web sites effectively, agricultural communicators have to take advantage of all of the tools at their disposal. There are many ways for creative people to get the word out about their site. In the marketing game, numbers rule. The more people who know about a Web site, the more successful it can be. The snazziest Web site cannot compete if people do not know about it.

Agricultural communicators must work hard to keep their sites current. In a fast-changing world, the public demands fresh content, and it is the responsibility of communicators to keep up with all of the changes.

The Internet and the ability to connect to other computers will continue to revolutionize agribusiness. Agricultural communicators who can keep up with the changes in the business world will stay successful for many years to come.

REVIEW QUESTIONS

Multiple Choice

1. An Internet Protocol (IP) address identifies every computer connected to the Internet by using

    a. a series of letters and numbers.
    b. a series of 0's and 1's.
    c. a series of letters.
    d. a series of numbers.

2. A resource used to help locate Web sites is called a(n)

    a. locator.
    b. infoseeker.
    c. search engine.
    d. locomotive.

3. One of the keys to a business's success is

    a. search engine ranking.
    b. assigned number.
    c. determined by what page it is on.
    d. none of the above

4. One of the most beneficial aspects of electronically connecting with computers is

    a. how distance is effectively closed between parties that link up.
    b. the speed and efficiency with which connection can be made.
    c. the ease of computer use.
    d. all of the above

5. The largest advantage of selling products on the Internet is the

   a. volume of sales.
   b. small overhead cost.
   c. store never closes.
   d. number of potential buyers reached.

6. If you get an e-mail from a business contact,

   a. never respond for fear the contact may think the reply is SPAM.
   b. respond only if you think it is a good contact.
   c. always respond.
   d. respond only after the second inquiry.

7. When designing a Web site, always use

   a. brightly colored backgrounds with (if possible) eye-catching graphics.
   b. basic colors, such as black or white.
   c. animation as much as possible.
   d. none of the above

8. Make navigation easy

   a. by using links that are no longer supported.
   b. by using several pages.
   c. with easy-to-read and -understand directions.
   d. none of the above

9. When designing a Web site, use

   a. Old English fonts that match the background.
   b. basic fonts that contrast with the background.
   c. small basic fonts.
   d. large block fonts that allow as few words on the screen as possible.

10. FTP stands for

    a. "for the people."
    b. "from the processor."
    c. "fast time processor."
    d. "file transfer protocol."

11. Marketing a Web site can be broken into _____ distinct segments.

    a. 1
    b. 2
    c. 3
    d. 4

12. A type of e-mail marketing is sending a message to

a. hundreds of prospective clients, known as shotgunning.
b. a targeted list of prospective clients.
c. a company that specializes in disseminating information to Internet users.
d. your friends and family.

13. Reciprocal links

a. are when someone puts your link on his or her Web site and you do likewise.
b. are the Golden Rule when pertaining to Web sites.
c. send users back to the home page.
d. allow a direct link to a user's PC.

14. Web rings are

a. Web sites joined in a sort of electronic marriage.
b. related Web sites that join to try to increase traffic for everyone within the ring.
c. both a and b
d. none of the above

15. In order for a Web site to be effective, it must

a. be unique.
b. be eye catching.
c. use provocative wording.
d. stay current.

Fill in the Blank

general	target	information
banner	two	to the point
covers	utilize	navigate
photos	mouse pads	link
read	relevant	general
simple	clip art	scenario
spelling		

1. Successful agriculture communicators must have the knowledge to _____ electronic communication in every possible _____.

2. An effective e-mail is _____, gets _____, and uses correct _____, and it always elicits a reply.

3. Two keys to successful Web sites are to make them easy to _____ and easy to _____.

4. When deciding on Web content, make sure all of the _____ that it _____ is _____ to the business.

5. Unless the Web site has a page that needs _____ or _____ to illustrate a concept or highlight products for sale, limit the artwork to about _____ per page.

6. _____ marketing is the process by which nonspecific groups are targeted.

7. _____ marketing involves aggressively seeking out people who are potential clients and telling them the Web address.

8. A _____ method of e-mail marketing is by adding a _____ to the site at the end of all e-mail messages.

9. One way to market using Web sites is by the use of _____ advertisements.

10. One way to promote a Web site is to give away _____ _____.

ENHANCEMENT ACTIVITIES

1. To discover the way search engines work differently, pick keywords, such as *beef cattle* or *soybean varieties*, and run them through sites like <http://www.google.com>, <http://www.alltheweb.com>, or <http://www.altavista.com>. When the results are returned, make a note of the top ten sites listed.

2. Print out a sample e-mail from your personal files, and try to identify an element that makes the e-mail ineffective as described by the text.

3. Find an agribusiness in your area that wants a Web site. Then consult with that business, and design and publish a site that highlights the products or services offered by the business.

4. Search the Internet for (1) agribusinesses that have effective Web sites as described by the text and (2) Web sites with ineffective designs.

5. After completing Activity 3, market the Web site using e-mail, Internet message boards, and other means available to you and the agribusiness. Track the amount of traffic that marketing the site generates by adding a counter to the site. You can find free Web counters by checking with sites such as <http://www.sitemeter.com> or <http://www.fxweb.com>.

6. After completing Activity 3, continue to keep the site's content fresh by consulting with the owner of the site often.

WEB LINKS

To find out more about the topics discussed in this chapter, visit these Web sites:

Common Search Engines
<http://www.google.com>
<http://www.alltheweb.com>
<http://www.altavista.com>
<http://www.excite.com>
<http://www.yahoo.com>

How Web servers and the Internet work
<http://www.howstuffworks.com/>

How E-mail Works
<http://www.howstuffworks.com/>

How Internet Search Engines Work
<http://www.howstuffworks.com/>

How E-commerce Works
<http://www.howstuffworks.com/>

How Web Pages Work
<http://www.howstuffworks.com/>

Alternatively, do your own search at <http://www.alltheweb.com>
(search terms: e-mail, Internet, e-commerce, file attachments, Web page design, Web site marketing)

BIBLIOGRAPHY

Sweeney, Susan. (2002). *101 Ways to Promote Your Web Site: Filled With Proven Internet Marketing Tips, Tools, Techniques, and Resources to Increase Your Web Site Traffic* (4th ed.). Gulf Breeze: Maximum.

Williams, Robin, and Tollett, John. (2000). *The Non-Designer's Web Book* (2nd ed.). Berkley: Peachpit.

CHAPTER 5

PHOTOGRAPHY: ILLUSTRATING THE STORY

OBJECTIVES

After completing the chapter, you should be able to:

- discuss the importance of photography to agricultural communications.
- compare and contrast film and digital cameras.
- discuss the basics of traditional film.
- discuss the basics of digital film.
- understand the basic rules of camera exposure.
- discuss the three basic rules of photography.
- explain tips for taking better photos.
- effectively illustrate a story.

KEY TERMS

aperture	ISO	rule of thirds
digital camera	lens	shutter speed
f-stop		

OVERVIEW

Photography is a universal language. A great picture can transcend language barriers, cultures, and even time. People have always had a need to record the world around them in pictures. Even prehistoric humans painted and scratched images of their world on cave walls.

For centuries, the only common way of recording everyday scenes was through painting, passing down stories verbally, or writing observations down on paper. It was not until the nineteenth century that photographic technology began its upswing (Figure 5-1) with the first permanent image, called the daguerreotype. A Frenchman named Louis Daguerre produced it in 1837.

Cameras were large and bulky, and they were not used by the general public. Exposure time for a single photo was an astounding half-hour. Photography, as both an art and science, crept along for more than 100 more years before any real breakthroughs began to show up on the marketplace. It was not until the 1950s that photographic technology became popular and widely available.

Since the 1950s, photography has undergone revolutionary changes in both the technological advances of the camera and the quality of film, as well as ease of use for consumers.

In 1985, photography saw yet another revolution when the world's first auto-focus 35mm single **lens** reflex camera was introduced. Some ten years later, another innovation, one that promises to change the way pictures are taken forever, began with the introduction of digital photographic technology. Digital cameras are a marriage of conventional film

FIGURE 5-1   It was not until the nineteenth century that photography began its upswing. (*Courtesy USDA/ARS*)

cameras with the electronic wizardry of microprocessors. (Digital cameras are discussed later in the chapter.)

Despite all of the advances—lens design, camera automation, film quality, and the like—the essence of photography has never changed. People will always look through the viewfinder of their camera, press the shutter release, and freeze a moment in time forever.

At its root, the word *photography* translates to "painting with light." For photographers, the film is their canvas, the camera and lenses are their brushes, and the light serves as paint. The urge to record images of families, friends, and occurrences all around us compel most of us to pick up a

FIGURE 5-2   Events must be recorded for the sake of our children. (*Courtesy USDA/ARS*)

camera from time to time. For the professional photographer or agricultural journalist, the urge to pick up a camera transforms from a luxury to a necessity.

News and events must be recorded for posterity (Figure 5-2). Imagine if we had no pictures of the Earth taken from space or no photographs of an elated child on her first birthday. It would indeed be an emptier world.

Photos tell us who we were and are. In a way, they help shape our future by documenting our greatest accomplishments in which we should bask. Yet photos also remind us of our shortcomings as a society.

This chapter examines many aspects of photography. The intent is to make you a better photographer by arming you with the information you need to make educated decisions about photographic equipment. In addition, this chapter provides some basic rules of photography that will make your pictures start to look more appealing instantly. The chapter ties all of the tips together by explaining how to illustrate a story in a professional and comprehensive manner.

FILM VERSUS DIGITAL CAMERAS

For the past 150 years or so, film cameras have been the mainstay of all photographic technology. Since its inception, film has undergone several refinements to ensure one thing: a better-quality image. Indeed, film has gotten better in many ways. Many formats, from the huge 16- by 20-inch sheet films of the early twentieth century to the super-small disc films of the 1980s, and everything in between, have been tried and have been successful from a technical standpoint.

The chemistry involved in making film has evolved to the point where film can accurately duplicate the range of color tones seen by the human eye in a variety of lighting situations. Although not as sensitive as the human eye, film continues to make the march in that direction.

For the most part, film was the standard medium in which to record an image until the mid- to late 1990s, when digital photography emerged on the heels of the super-fast home computer and the Internet. For the first time in its brief history, film saw its destined successor in the world of photography.

Digital photography is an extremely versatile, easy-to-use medium, and has a variety of applications for agricultural communicators. As the technology increases in quality and speed, there is no doubt that **digital cameras** will be the norm and film cameras will become curious artifacts that only photographic purists use.

Despite their differences, both types of photographic formats have similarities.

Both Can Be Extremely Easy to Use

Today's cameras are made specifically for ease of use. In a culture of quick fixes, few people want to bog themselves down with technical details on how to fine-tune the settings on their camera. Instead, they want to be able to shoot pictures right after taking their new camera out of the box.

With point-and-shoot simplicity, today's moderately priced film and digital format cameras come with a shallow learning curve. Most people can master the technical functions of them relatively quickly.

Both Are Readily Available

With the advent of the Internet and on-line commerce, the latest in camera technology is just a few mouse clicks away, no matter how remote the place where you live. Nearly every major department store retailer carries both digital and film format cameras for the ultimate in choice for the consumer. Trying one out is as easy as a trip to the local shopping mall.

Both Are Relatively Inexpensive

Would you believe that you can buy a 35mm film camera for under ten dollars (Figure 5-3)? And what would you think about purchasing a new digital camera for under twenty dollars? Both are available. Today's ultra-competitive consumer markets have driven prices on electronic units

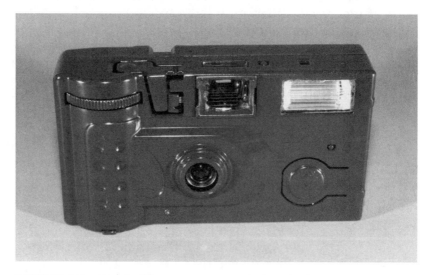

FIGURE 5-3   Simple 35mm cameras can be bought for less than ten dollars. (*Courtesy Russell A. Graves*)

lower and lower. Although the durability of the low-end units and the quality of the images they produce leave something to be desired, they nonetheless can produce an image.

For around $300, you can purchase a decent film or digital camera that will work well in a variety of situations. Nevertheless, with camera prices, a direct correlation exists between the price of the unit and its overall quality. In short, you get what you pay for.

DIGITAL CAMERAS

Digital cameras are a marriage between conventional film cameras and computer technology. This marriage of the optical qualities of a traditional camera coupled with state-of-the-art digital imaging technology has produced a photographic hybrid.

Instead of using film, the digital camera uses semiconductors and light-sensitive computer chips to record a photographic image. When light strikes the computer chips (known in technical terms as a charged coupled device, or CCD), the image is temporarily "memorized" on the CCD. The semiconductors in the camera process the image into a binary code and store the photo on either a floppy disk or some other type of memory storage device for later retrieval.

Digital cameras transform images into digital form, where they can be stored on electronic media such as CDs and disks.

As a rule, the number of pixels contained on the CCD governs the quality of image that a digital camera can produce. Lower-end digital cameras typically have a 640 × 480-pixel count and can produce acceptable prints up to 3 by 4 inches in size. High-end digital cameras have pixel counts of 2000 × 1500 pixels and can produce quality images up to 8 by 10 inches in size. The higher the pixel count, the more a camera will cost.

Once the images are retrieved from the memory storage device via a computer, they can be manipulated in a variety of ways, including cropping and color correcting, all within the computer. Once the image is digitally processed, it can be e-mailed, posted on a Web page, archived to a compact disc or other types of digital storage, or printed by a high-quality ink-jet printer.

Digital Cameras

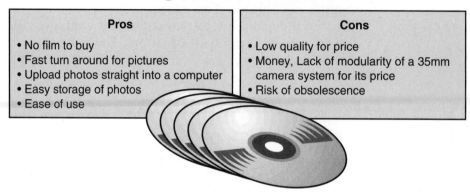

Pros	Cons
• No film to buy • Fast turn around for pictures • Upload photos straight into a computer • Easy storage of photos • Ease of use	• Low quality for price • Money, Lack of modularity of a 35mm camera system for its price • Risk of obsolescence

FIGURE 5-4   Pros and cons of digital cameras.

Digital cameras have lots of positives but also some negatives (Figure 5-4).

Pros of Digital Cameras

No film to buy   Digital cameras capture and then store each image on a 3.5-inch floppy disk (Figure 5-5), memory cards that slide out of the camera and into printers or computers, or an internal hard drive inside the camera and transfer the images via a cable.

Because all of the images are stored on digital media, film is not needed. What are needed are extra memory cards or floppy disks if you plan to do a lot of shooting since computer memory is finite. With current film prices around $3.50 a roll for a 24-exposure roll of Kodak 400 speed film, plus about $5.00 for processing, the cost incurred on a roll of film runs close to $10.00.

For digital images, the overall cost of producing an image is much lower since digital memory can be used repeatedly compared to the single-use function of film. Once photos are downloaded on a computer, they can be analyzed and discarded if need be.

Fast turnaround for pictures   Perhaps one of the main advantages of digital photography is the ability to review pictures merely seconds after they are taken. Unlike traditional photography, which requires taking film to a lab for processing, digital images are ready the instant they are taken.

FIGURE 5-5 Digital cameras can store photos on floppy disks and CDs. (*Courtesy Russell A. Graves*)

Although advances in film processing have sped up the developing times to under an hour, drive time, as well as other variables, can slow the process of getting pictures in your hand to some degree.

Because of the speed and efficiency with which digital images are ready for viewing, many newspapers and current event Web sites employ the use of digital photography. Because of their ability to record late-breaking news and transmit the photographs to a central news bureau via the Internet in a matter of minutes after the event occurs, the news reporting loop from occurrence to reporting tightens even more.

Upload photos straight into a computer   With traditional film photography, images have to be scanned off the film or the finished pictures have to be scanned in order to get the images on a computer for digital editing or archiving. With digital cameras, the images are uploaded to a computer—skipping the scanning process and ultimately saving the user time.

Cameras use memory cards, 3½-inch floppy disks, or high-speed Universal Serial Bus (USB) connections to transfer photos from camera to computer. With the acceleration of computer technology, wireless transfer of photos from camera to computer is no doubt right around the corner.

Easy storage of photos   If the digital images are kept, the process of cataloguing them and storing them on digital media for permanent archiving is extremely easy and cost-efficient. A single compact disc can store around 600 jpeg images scanned at a resolution of 1350 pixels per inch for about a dollar per compact disk. A jpeg image is a commonly used file type that compresses digital image information to keep file sizes relatively small.

When the images are stored digitally, they can be easily catalogued using keyword descriptions or the dates the photos were taken, and any technical data about the photograph can easily be stored in the digital realm along with the image. Once an archival database is in place, finding photos is a matter of entering a few keystrokes with the picture's description.

Currently, finding photos, especially in a large, conventional photo library, is a time-consuming task that requires the photo researcher to look through hundreds, and maybe even thousands, of images to locate a few select images.

Ease of use   Today's consumers demand that electronic equipment be user friendly: once the product is taken out of the box, the new owner wants to have results within a matter of minutes. This principle applies to most electronic items, including digital cameras.

In the current market, three digital camera price ranges exist: under $200, $200 to $800, and $800 and above. Most consumers are shopping in the first two groups. The low- to midrange-priced cameras feature point-and-shoot simplicity and are extremely easy for novice photographers to use.

At the upper end of the digital camera market, the cameras become a little harder to use for the novice photographer. Packed with features designed for professionals, the upper-end digital cameras can be daunting for the new photographer to tackle. Nevertheless, with a little patience and an instruction manual, novices can master even the most complicated of cameras.

Cons of Digital Cameras

Lower quality for price   For the same money, digital cameras cannot compete with film in terms of quality. A $500 film camera can take much higher-resolution photos than a digital camera of the same price.

The reason? Like any other kind of electronics, new technology is expensive. Once the technology has saturated the consumer market, prices start to fall. Just a few years ago, digital video disk (DVD) players were over $1,000 when purchased new. Today, a quality DVD player is readily available for under $200.

In the year 2000, top-quality digital single-lens reflex cameras cost more than $30,000. Currently, cameras capable of fine resolution are available for $4,000 to $5,000, but even at that price, they still are not as good as a 35mm film camera.

For now, though, if high-quality images for a reasonable price is your goal, you cannot beat a slow-speed, finely grained 35mm film.

Less modularity for price   For the same money, digital lacks the modularity of 35mm camera systems. Digital cameras under $800 are typically of the point-and-shoot variety. That is, you point where you want to take the picture and push the button; there is no fiddling with camera settings or adjusting lenses. Most cameras have a flash, but often, the camera decides when the flash should be used. In short, many low- to midrange digital cameras lack creative controls that advanced photographers can employ.

Higher-end digital cameras are capable of changing from wide-angle to telephoto lenses to fit a particular photographic challenge. Many camera manufacturers, such as Nikon, have dozens of lenses available for their digital cameras that accommodate a variety of situations that a photographer may face.

In addition to lenses, other extras include flashes and full creative controls, such as the ability to set **aperture** or **f-stop**, and **shutter speeds** to match a situation. They are available for those willing to spend the money on the best that the digital world has to offer.

Any midrange 35mm single-lens reflex camera enjoys wide latitude of modularity for around $200.

Risk of obsolescence   Since digital cameras are a product of the computer age, their obsolescence can come quickly. Technological research, manufacturing processes, and the speed with which semiconductor innovations are taking place are astoundingly fast. The computational power of computers has grown exponentially. Microchips have grown smaller and smaller, and their computational powers have become faster. Computers that were the size of small cars thirty years have been shrunk to the size of small calculators and wristwatches.

Digital photographic technology is evolving fast. What is state of the art today may be common technology a year from now. With technology moving at such a fast pace, it is unsure whether we will reach a technological plateau. For now, however, the evolutionary pace continues to quicken and digital cameras continue to improve and drop in price, making a camera bought today technologically obsolete tomorrow.

FILM CAMERAS

With the great advances on the digital photography front, it would seem that film cameras are dinosaurs. Right? Nothing is further from the truth.

Film cameras are still extraordinary values for those who want to take high-quality images at a reasonable price. Even the cheapest 35mm cameras, so long as they are equipped with good lenses, will still outdo digital cameras costing ten times more when a comparison is made in terms of image quality (Figure 5-6).

Film cameras accept a wide array of film formats, such as color slide, color print film, black and white film, or infrared (Figure 5-7). Depending on the lighting conditions, film can also be matched to a variety of situations from indoors under artificial light to full sunlight or dimly lit interiors.

FIGURE 5-6   Film still offers superior results compared to digital imaging. (*Courtesy Russell A. Graves*)

FIGURE 5-7   Film cameras offer a wide variety of film types. (*Courtesy Russell A. Graves*)

No other camera exists today that combines a compact design, comprehensive modularity, and affordability as today's 35mm single-lens reflex camera. Like digital cameras, film cameras have both pros and cons associated with them. Here is a run-down of each (Figure 5-8).

Film Cameras

Pros	Cons
• Ease of use • Many models at a relatively inexpensive price • Readily available film rrocessing • A variety of accessories	• Complex features to master • The need to scan to put pictures on a computer • An ongoing investment in film and processing

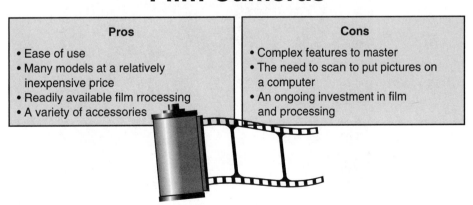

FIGURE 5-8   Pros and cons of film cameras.

Pros of Film Cameras

Ease of use   From the least expensive to the most expensive, film cameras, with the possible exception of those on the extreme upper end in price, are extremely easy to use. Consumers have a choice when buying the simplest of film cameras: the 35mm disposable camera, which works well in a variety of situations. For a modest investment of about five dollars per camera, everyone at a party can have a camera to shoot their perspective on the event.

You also have the choice to buy at the high end of the 35mm spectrum (Figure 5-9). Somewhere in between lies the realm of the 35mm point-and-shoot and low- to midrange 35mm single-lens reflex (SLR) cameras. Point-

FIGURE 5-9   As a consumer, you have a choice when buying high-end equipment. (*Courtesy USDA/ARS*)

and-shoot and SLR cameras are extremely easy to master and use in a relatively short length of time.

Many models at a relatively inexpensive price   One major camera manufacturer, Minolta, manufactures eight different models of 35mm SLR cameras at an average price of around $400. If you take away the most expensive and least expensive camera from that bunch, the price drops to an average cost of $289 per model.

The quality of construction from the high end to the low end of the cameras varies, and the most expensive of the eight models have many professional features not found in the lower-end models. There is nevertheless a common denominator that weaves through each of the camera models. If you put the same lens on each of the cameras and took a photo of the same subject under duplicate conditions, you could not tell which camera took which image.

Whether it is Minolta or any other major camera manufacturer, the quality of the image that each camera can take is identical from one model to the next. It is lenses and film selection, not the camera, that ultimately influences the quality of an image. (There is more on that subject later in the chapter.)

Readily available film and processing   One of the features of film cameras, and a definite advantage, are all the varieties of film for it. Discount department stores are well stocked with films in a variety of brands and speeds. Not all films are created equal. Just like brands of automobiles are different in a number of ways, so is film. Brand name and film speed are the two biggest factors that influence the way an image appears.

Different brands of film, such as Kodak, Fuji, and Agfa, render colors differently. For example, Fuji films have deeper blues and greens in their photos than do Kodak. That is not an endorsement of Fuji films; rather, it is a suggestion to experiment with different types of films and determine what brands suit your needs under various conditions.

Film speed is also a consideration. Various film speeds are meant to be used under different lighting circumstances (Figure 5-10). It is best to experiment and use what works best for you. (Later in the chapter, we discuss film speed in more detail.)

Just as film is easy to find, so is film processing. Many national pharmacy chains offer photo processing as one of their services, as do discount department stores. Prepaid film processing mailers can be purchased. The exposed film is mailed to a processor and returned by mail.

Processing times can vary from under an hour to a week or more if you are using slide film or prepaid film mailers. Whatever the wait, film processing is still easy to find.

FIGURE 5-10   Different films are meant to be used under different lighting conditions. (*Courtesy Russell A. Graves*)

A variety of accessories   The current line of Canon photo products features twelve SLR cameras and a complement of more than fifty lenses and thirteen flashes that fit any of the cameras Canon makes.

The true advantage of 35mm SLR cameras is their modularity (Figure 5-11). You can outfit a camera with remote controls and fire them from hundreds of feet away, or buy lenses that focus up to fractions of an inch away from a subject, or zoom in to an object a hundred feet away. The way a camera is outfitted is ultimately limited only by budget and imagination.

Cons of Film Cameras

Complex features to master   Like digital camera, film cameras, especially the ones loaded with features and wedged in the upper price brackets, can be intimidating for the novice and even the experienced user. *Depth of field preview, exposure compensation, multiple exposure capabilities, mirror lock-up:* What do they mean?

Many people view owner's manuals as a bore to read. However, they are essential for learning a camera's capabilities. Unfortunately, for each minute you spend reading, that is a minute less you have to learn the essential functions of your camera's operations. For seldom-used camera functions, they can be especially daunting to learn and remember. When news starts breaking fast, you may not have time to read about how your camera works.

FIGURE 5-11   The true advantage of 35mm cameras is their modularity. This feature allows a wide range of specialized equipment to be used, such as macro or close-up lenses. (*Courtesy Russell A. Graves*)

The need to scan to put pictures on a computer   Unlike digital cameras, the pictures from a film camera must be manually scanned in order to manipulate them in a digital darkroom (Figure 5-12). This may not be a problem since both flatbed and film scanners are very affordable and easy to use. However, since an extra step is needed to produce a final photographic product, it does take time, and time is money.

FIGURE 5-12   To put traditional pictures on a computer, they must be scanned. (*Courtesy Russell A. Graves*)

It takes a minute to scan a photograph and a minute to manipulate it digitally, so the total investment is two minutes per picture, which does not sound like a lot of time. But many photo assignments generate several hundred photos. If you had to scan every photo in a one-hundred-image presentation, then the total time invested in the final stages of the project would be 3 hours and 20 minutes of just scanning and manipulating. And that does not include the time to load new pictures on the scanner or run the keyboard and mouse. If your time is worth $10 an hour, you will lose about $50 per assignment. Ten or more assignments, and you can see the kind of money that manual scanning costs.

An ongoing investment in film and processing   If you take pictures, you know that you have to buy film and pay to have it processed. A roll of film, by the time it is processed, can cost about $10. Many pros shoot one hundred or more rolls of film on an assignment, so it is clear how quickly the costs of a project can add up.

With digital cameras, the memory can be used repeatedly; and in essence, you have an endless supply of renewable film. With traditional film cameras, once you push the button, that frame of film is altered forever. You may have gotten a good picture or a poor one. Once the shutter is tripped, it does not matter because the image is captured for better or worse.

Many experienced photographers advise being judicious with the use of film. However, in photography, quality always beats quantity. Learn to master your camera and the art of taking pictures; then burn up the film.

THE THREE BASIC RULES OF PHOTOGRAPHY

Many variables are involved in taking quality photos. Everything from lens selection to camera handling can spell the difference between a picture that you are proud to show off and one that goes in the trash.

Today's cameras are technological marvels. They are extremely adept at handling all sorts of lighting situations, they can auto-focus on a subject, and they load and wind the film automatically. In addition, most modern 35mm SLR cameras automatically set the aperture and shutter speed according to the available light. These types of camera make photography virtually foolproof in the hands of inexperienced users.

The three rules of photography:
- **Get close**
- **Keep the sun at your back**
- **Use the rule of thirds**

With shutter speed, focus, and aperture settings a given, here are three tips that can help you make the jump from mediocre photos to memorable ones. These rules, like so many others in photography, are not hard and fast. They are merely general guidelines in which to go by in order to start achieving more consistent and predictable results. A good way to substantiate these rules is to look at magazines that have many photos in them.

You will find that the vast majority of photos were taken according to these three basic rules.

Rule No. 1: Get Close

One of the biggest mistakes that beginning photographers make is not getting close enough to their subject. Many times a photo is composed and shot, only to reveal disappointment when the processed film comes from the lab. The subject that was clearly visible in the viewfinder is just a speck on the finished picture. The reason for this is a commonly held misconception that with a big lens, a photographer is able to stand a mile away from the subject and take an acceptable photo. Nothing can be further from the truth. In fact, in order to get an acceptable picture of the subject, you have to move yourself closer to avoid the dreaded "speck" syndrome.

You can get closer to your subject in two ways: optically and physically. Getting closer to the subject from an optical standpoint is easy to achieve. All it takes is using a telephoto lens. In photography terms, a telephoto lens is anything that magnifies normal vision. Normal vision is a 50mm lens. Anything above that is considered telephoto. Telephotos come in a variety of ranges, from 70mm to 1000mm, with 200mm (four times normal view) being the most common.

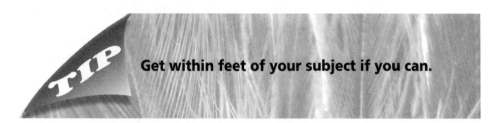

Get within feet of your subject if you can.

Getting physically closer is the easier and less costly alternative for abiding by Rule No. 1. Getting physically closer means moving your body closer to the subject. For a person, that is relatively easy, but for shy subjects such as wildlife, a telephoto lens should be used. A good rule of thumb is to try to get as close as you can and make the main subject fill most of the frame that you are composing.

Rule No. 2: Keep the Sun at Your Back

This rule is especially important. By keeping the sun at your back, your subject will be more evenly illuminated by natural light, which is far superior to any artificial light, such as electronic flashes or indoor lighting.

Keeping the sun at your back makes shadows fall away from your subject.

As a rule, keeping the sun at your back cannot be accomplished during midday because the sun is overhead. Midday light is hard for shooting good photos because the sun is shining straight down on the subject and lighting the top of it. Consequently, the best time to shoot photos in terms of the quality of light is before 10 A.M. and after 4:30 P.M.

During these times of optimum photographic illumination, the light takes on a softer and more reddish appearance that lends itself well to outdoor photography. Skin tones take on a more natural appearance, colors appear richer and more natural on film, and the overall quality of the photo is enhanced.

Do not lock yourself into keeping the sun at your back all of the time, however. If the sun is low enough, your shadow will appear in the photo, causing a less desirable image. In these cases, move around the subject and try side lighting or backlighting.

This rule does not apply when shooting silhouettes. It should be noted however, that when shooting silhouettes, you should not look at the sun through the lens of a camera for more than a brief instance.

Rule No. 3: Use the Rule of Thirds

The **rule of thirds** refers to the composition of a photo. The brain finds that photos that are composed a little off center are more appealing than ones that are centered. Using the rule of thirds is simple. Imagine that your viewfinder is broken up into even thirds across the horizontal and vertical planes. Wherever the lines intersect is where the main emphasis of the photo should be located (Figure 5-13). For example, if you are shooting a photo of a landscape, the horizon line should run along either the lower third or upper third of the photo, not right in the middle. In other words, do not center your photos.

Play around with the composition until you find something that is appealing. The same rule applies to a vertical composition. Many photographers find themselves locked into horizontal and cannot see the world otherwise. Try to take photos in both the horizontal and vertical formats, remembering to use the rule of thirds, and see which ones look the best. Film is cheap, and the only way to get better is to practice.

FIGURE 5-13   The subject should be placed close to where the lines intersect. (*Courtesy Russell A. Graves*)

If you can master this one rule, your photos will immediately start to look more appealing.

RULES OF THE ROAD: HOW TO ILLUSTRATE A STORY AND OTHER HINTS

Step 1: Picking the Right Film

One of the most important decisions concerns the kind of film to use. The selection process can be daunting. Many different brands, emulsions, and film speeds can make choosing film downright intimidating. A systematic approach to picking the right film is the key.

The first decision is whether to use print or slide film. Print film is the most common type of film used by consumers and comes in a variety of speeds. The **ISO** number, as it is called, gives an indication of the film's sensitivity to light; the number is written prominently on boxes of film. The system for determining which type of film to use is simple: The higher the film's ISO number, the more sensitive it is to light. For example, ISO 1000 speed film can be used in lower light conditions than ISO 100 speed film.

A good rule of thumb to go by is to use the lowest-speed film possible. As the film speed gets higher, the quality of the image suffers. A comparison of an 8 by 10 inch enlargement taken with ISO 1000 film to that taken with ISO 100 film would show that the first enlargement is grainier than the one made from lower-speed film. The slower-speed film produces crisper images with a tighter grain when reproduced (Figure 5-14).

Slow films with low ISO numbers are not practical in all situations. For indoor shots or low-light shots, use a fast film such as ISO 1000. For photos in bright sunlight, ISO 100 is the best. Many people believe that ISO 400 is the best all-around film. It works well in sunlight and dimly lit situations. Although a little grainier, the images are still of good quality when enlarged.

On the subject of slide film versus print film, the differences in the two are found in the quality of the first-generation image. When you use slide film and the image is exposed, the film produces a positive image, with all of the colors true to life on the piece of film. Once the roll is developed, the actual piece of film that the picture is on is cut and mounted on cardboard or plastic mount. The image in the mount is a first-generation image and usually has superior color and contrast (Figure 5-15).

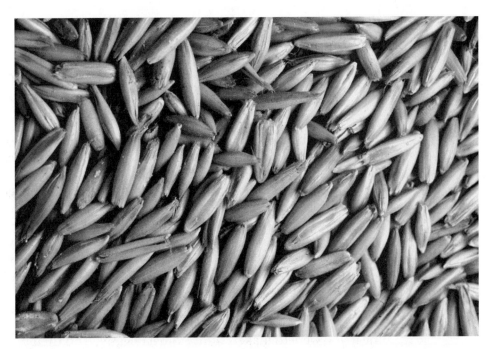

FIGURE 5-14   Slower-speed films produce crisper images. (*Courtesy Russell A. Graves*)

FIGURE 5-15   A slide is a first-generation image. (*Courtesy Russell A. Graves*)

Most publishers of books and magazines prefer using slide film to reproduce images in their publications. The image reproduces much more true to the image than print film does.

A print made from print film is a second-generation image (Figure 5-16). When print film is exposed to light, a negative image is produced on the film. On the negative, light parts of the image appear dark, and dark parts appear light. The negative is the first-generation image, yet as a photographic record, it does us no good because of the juxtaposition of colors. Only when a print is made from a negative can we fully comprehend the contents of a photograph.

A photograph is a second-generation image. Each time an image is produced from another image, the quality is compromised to a degree. Imagine if you were to make a print from a negative and then make a print from that print. This third-generation image would not be nearly as sharp as the first generation.

FIGURE 5-16   Prints are second-generation images. (*Courtesy Russell A. Graves*)

Print films do have a place for the agricultural communicator. The prints are easy to scan into a computer, and the film is typically less expensive than slide film and can be processed in many more places.

Step 2: Choosing the Right Equipment

The right camera equipment to use for a given situation is a source of debate among photographers everywhere. The brand of equipment and the type of equipment you use are largely personal preferences, but there is a starting point. For clarity, all of the equipment discussed in this chapter is 35mm SLR format since this is overwhelmingly the format of choice by agricultural communicators.

A good all-around photo equipment package consists of a camera body, a 28–80mm zoom, an 80–200mm zoom, and a flash. This camera package does not cost a lot, and all the components are easy to find.

All SLR camera lenses measure their focal length in millimeters. *Focal length* is the distance from the primary lens element to the film plane (Figure 5-17). This is often a source of confusion for novice photographers, but it is very easy to figure out.

FIGURE 5-17   Focal length is the distance from the lens element to the film plane. (*Courtesy Russell A. Graves*)

Humans see in what is roughly equivalent to a lens that is 50mm in focal length. Lenses around the 50mm are called "normal" lens in photographic circles. Anything smaller than 50mm—say, a 28mm—is considered a wide-angle lens because it takes in a wider view than a normal lens does (Figure 5-18). Any focal length above 50mm is a telephoto. For example, a 200mm lens is equivalent to human vision magnified four times.

A common mistake that people who have a budget for photo equipment make is they buy the best camera they can afford and skimp on lenses. This is a backward approach to ensuring quality photos because lenses influence the quality of an image much more than the camera does. Cheap lenses produce substandard photographic results in the form of fuzzy pictures and colors that are not true to life. If you have to skimp on something, buy a less expensive camera, but spare no expense in purchasing lenses.

Camera flashes are another source of misunderstanding with novice photographers. If you have ever been to a professional sporting event or concert, you will see people high up in the stands using flashes to shoot pictures of a subject that is a hundred feet away. For effective flash photography, this practice is a waste of batteries.

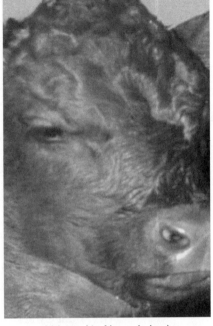

50mm (Normal view)   200mm (4× Normal view)

FIGURE 5-18   By changing lenses, a photographer can drastically change the perspective. (*Courtesy Russell A. Graves*)

For most flashes, try to stay within 20 feet of the subject so that you can take advantage of the light that the flash produces for great photographic results. If you go beyond the 20-foot limit, the flash is not powerful enough to illuminate the subject properly.

A camera's flash is good only to a little over 20 feet away from a subject.

Step 3: Get to Know Your Subject

Whenever that first photo assignment comes your way, you need to be ready. The first step in being ready is to learn your subject thoroughly.

Whether it is a person or an inanimate object, an intimate knowledge of your subject is vital.

Start by doing background research of the subject. If the subject is a person, try to find out what the person's hobbies are or find a subject in which he or she is deeply interested. Since most people are a bit camera shy, it is the photographer's job to help them relax and be at ease.

When you have some background information about the subject, two things happen. The first is that you can photograph the subject in a pose or setting that transcends the ordinary. Maybe the subject is the county's top farmer. A photo of the person taken from an elevated vantage point standing amid a large field of cotton and looking up at the camera would be much more interesting than a traditional eye-level shot.

When you get to know your subject, you become a director of the scene instead of a participant. If you have taken the time to research your subject, interesting conversation will guide the photo session, and the subject will relax. The result will be photos that are more natural.

If your subjects are not people, research can still play a big role in the success of an assignment. If you were hired to shoot pictures of external parasites of cattle, research would prepare you to tell the difference between face flies, heel flies, horn flies, and horseflies (Figure 5-19). Without research, you might not shoot the right flies.

FIGURE 5-19   With research, you will be able to identify this insect as a horsefly. (*Courtesy Russell A. Graves*)

Step 4: Look at the Big Picture

When it is time to shoot your first assignment, take the time to study your surroundings. Consider the big picture (Figure 5-20). What do you see? What looks interesting? What are some effective backgrounds for pictures? Ask yourself these questions when you are analyzing what you want to take a picture of.

Often, it is helpful to go to the location a day or two ahead of time and scout the area for photographic possibilities. That way, just like any other job you may tackle, you are prepared for the challenge that lies ahead.

No doubt, you have heard the saying, "You can't see the forest for the trees." It applies to photography too. If you are so bogged down in looking at one little piece of the puzzle, you may miss the whole picture. When you approach a subject, step back and look at the whole scene. Then decide what pictures you want to take.

Step 5: Details, Details, Details

Once you have looked at the big picture and shot scenes of vast fields or pastures full of cattle, it is time to slow down even more and look for details that will enhance your photographic coverage of the subject (Figure 5-21).

What about the subject provides an interesting detail that will add a touch of visual flair to your portfolio of pictures on the subject? If you are shooting photos for an article about mad cow disease, what could you add to the package to make the photos create more of an impact?

You could take pictures of hundreds of cattle in a feedlot or even a row of steers with their head in a trough. That is the big picture. What about

FIGURE 5-20   When you start to photograph a subject, look at the big picture. (*Courtesy Russell A. Graves*)

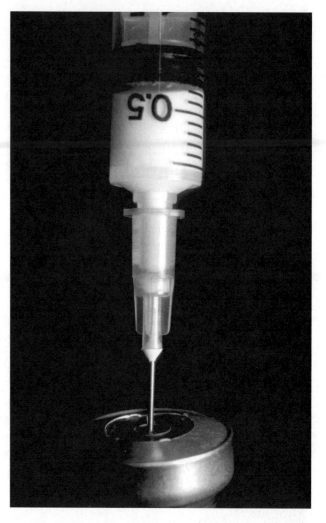

FIGURE 5-21   Always look for details in a subject. (*Courtesy Russell A. Graves*)

the details? Consider a close-up of a steer's head as he stares at you. Or a photo of the feed the cattle eat, where the grain takes up the whole frame of the viewfinder. Maybe a tight shot of a cowboy's hand injecting an antibiotic in an animal with a syringe would work.

Details, details, details. Take the time to look for them.

Step 6: Editing Your Photos

You have followed all of the rules presented in this chapter for taking pictures. You have the best lenses you can afford, and you shot pictures by looking at the big picture and then paying attention to the details. You have the photos back from the lab, and now it is time to edit them.

Start with the easy decisions. Look at photo editing as a job that is to be done in phases. The first phase is to look through the photos and cull the ones that are not technically perfect. Flat, uninteresting lighting, poor exposure settings, and blurred pictures are all candidates for early entry into the trash bin.

After the initial edit, phase two begins. Now look for pictures that have interesting angles and great lighting, and that follow all of the rules of composition and getting close to the subject. Anything that does not pass that test should not necessarily be thrown away. Just set these photos aside for other uses—perhaps later or on another similar assignment.

The third and final phase of the editing process is refining the images to about forty of the best that you have. This is perhaps the hardest phase to conduct. If sixty images passed the first two phases, how do you cut even more? First, look for duplicate images—maybe not exact duplicates but ones that look enough alike that one can be easily eliminated without disrupting the visual flow of the work.

Next, look for pictures that disrupt the continuity of the body of work. For example, if you have been taking photos of a farmer who is known for using innovative global positioning technology, you may have a great shot of a sunset that evening. Although the sunset shot is a great image, it does not have anything to do with the subject at hand.

If you still have not culled enough photos, go back to phase one and make sure that all of the photos that are not technically excellent have been eliminated. If that does not work, get ruthless. Edit any image that does not catch your eye, and force yourself to cull the images even more.

Editing can be a real chore, but it is also very rewarding when you see all of your hard work in the field coming together in a cohesive, artistic way.

CONCLUSION

The art of visual communication is important to the field of agricultural communications. From the earliest days of photography, people have sought to use the medium as a reliable and convenient way to record the events of our lives in a fashion that is accurate beyond reproach.

For the photographer, the ability to use equipment in a professional and technically perfect fashion is in great demand in many areas of the editorial and advertising markets.

Like any other discipline, photography is a skill that is developed over time. It takes practice to perfect the skill. That practice involves having a camera, and being ready to use it at all times.

The best way to start the path down the road to successful photography is to start with the basics. This chapter set out to provide a good foundation on the basics of photography and how to start taking pictures successfully. In addition, it provided a fair assessment of the pros and cons of both digital and film cameras so that you can make a wise decision concerning these two alternatives

Photography is many things. It is a way to record the moments of your life and others around you. It is a rewarding hobby and can be a lucrative career. The telling factor is how much time you are willing to learn and practice the craft.

REVIEW QUESTIONS

Multiple Choice

1. Sheet film from the early twentieth century was
   a. 10 inches by 20 inches.
   b. 12 inches by 20 inches.
   c. 14 inches by 20 inches.
   d. 16 inches by 20 inches.

2. If light strikes the charged coupled device,
   a. it is ruined.
   b. it holds a positive charge until a negative potential is felt at the diode.
   c. the image is temporarily memorized.
   d. none of the above

3. The quality of image that a digital camera can produce is determined by
   a. the number of pixels.
   b. the price.
   c. how many pictures it can store.
   d. the manufacturer.

4. To put pictures on a computer, you have to have them
   a. digitally mixed.
   b. scanned.

    c. on slide film.

    d. none of the above

5. The two ways to get closer to your subject are

    a. telescopically and optically.

    b. telescopically and physically.

    c. optically and physically.

    d. all of the above

6. A telephoto lens is anything that magnifies

    a. telephone signals.

    b. 100 times.

    c. 200 times.

    d. above normal vision.

7. The best time to shoot photos in terms of quality of light is

    a. after 10:00 A.M. and before 4:30 P.M.

    b. before 10:00 A.M. and 4:30 P.M.

    c. after 10:00 A.M. and 4:30 P.M.

    d. before 10:00 A.M. and after 4:30 P.M.

8. The rule of thirds is to

    a. use a different lens for each third of your pictures.

    b. use a different camera for each third of your pictures.

    c. imagine your viewfinder is divided into thirds horizontally and vertically.

    d. none of the above

9. The higher the film's ISO number,

    a. the more exposures it has.

    b. the less sensitive it is to light.

    c. the more sensitive it is to light.

    d. all of the above

10. Slower speed film produces

    a. crisper images with tighter grain.

    b. crisper images with tighter grain only if a flash is used.

    c. photos that show motion without blurring.

    d. duller images and looser grain when no flash is used.

11. The best all-around film is probably

    a. ISO 100.

    b. ISO 200.

    c. ISO 300.

    d. ISO 400.

12. A good all-around photo equipment package consists of a
    a. camera body, 28-80 mm zoom, 80-200 mm zoom, and flash.
    b. camera body, 50 mm lens, 300 mm zoom, and flash.
    c. camera body, 50 mm lens, and flash.
    d. none of above

Fill in the Blank

50mm	reddish	superior
accurately	slide film	ISO 1000
color print	versatile	composition
computer	lens	situations
emphasis	softer	conventional
first generation	duplicate	outdoor
indoor	infrared	reproduce
interesting	color slide	applications
lighting	practice	medium
marriage		

1. Film can _____ _____ the range of color tones seen by the human eye in a variety of _____ _____.

2. Digital photography is an extremely _____, easy-to-use _____ that has a variety of _____.

3. Digital cameras are a _____ between _____ film cameras and _____ technology.

4. Film cameras accept a wide array of formats such as _____, _____ film, black and white film, or _____.

5. Normal vision is a _____ _____.

6. During times of optimum photographic illumination, light takes on a _____ and more _____ appearance that lends itself well to _____ photography.

7. Wherever the lines intersect is where the main _____ of the photo should be located.

8. Film is cheap, and the only way to get better is to _____.

9. For _____ shots or low-light shots, use a fast film such as _____.

10. A _____ image usually has _____ color and contrast.

11. Books and magazines prefer using _____ to _____ images in their publications.

12. During the editing process, look for pictures that have _____ angles and great lighting, and follow all of the rules of _____.

ENHANCEMENT ACTIVITIES

1. Compare the quality of digital photos and still photos and see if there is any discernible difference.

2. Create a portfolio of magazine photographs that use the three basic rules of photography as described in the text.

3. Research brands of 35mm camera systems. Search the Internet and find Web sites by professional photographers. E-mail the photographers, and see if you can find out the brand and the type of camera systems that they use.

4. Pick an agricultural subject to photograph. Create a photo essay consisting of twenty photos concentrating on the big picture and the details. While working on the essay, take a roll of at least thirty-six pictures. Once the photos are developed, edit your work to the twenty best photos.

WEB LINKS

To find out more about the topics discussed in this chapter, visit these Web sites:

How Digital Cameras Work
<http://www.howstuffworks.com/>
  (Search term: digital-camera)

How Cameras Work
<http://www.howstuffworks.com/>
  (Search term: camera)

Film or Digital—What is Right for You?
<http://www.yls.org/>
  (Search term: techtalk)

Taking Great Pictures
<http://www.kodak.com/>
  (Search term: taking pictures)

How Film Works
<http://www.kodak.com/>
   (Search term: film basics)

Guide to Better Pictures
<http://www.kodak.com/>
   (Search term: picture taking)

Alternatively, do your own search at <http://www.alltheweb.com>
   (Search terms: digital photography, agricultural photography, film,
      photography, photojournalism, photo editing)

BIBLIOGRAPHY

Burian, Peter K., and Caputo, Robert. (1999). *National Geographic Photography Field Guide: Secrets to Making Great Pictures*. Washington, D.C.: National Geographic.

Instructional Materials Service. (2002). *Agriculture Communications 315 Course Materials*. College Station, TX: Author.

Shaw, John. (1984). *The Nature Photographer's Complete Guide to Professional Field Techniques*. New York: Amphoto.

CHAPTER **6**

SPEAKING FOR THE PUBLIC

OBJECTIVES

After completing this chapter, you should be able to:

- plan and prepare public speeches.
- understand the steps in preparing extemporaneous speeches.
- make the most of your time in the public eye.
- understand the role of a spokesperson.
- answer impromptu questioning.
- assemble illustrated talks.

KEY TERMS

extemporaneous speaking   spokesperson
prepared public speaking

OVERVIEW

For many people, speaking in public is a gift or a curse. A survey was conducted many years ago about what people feared the most. Surprisingly, most people responded that they have a fear of speaking before a crowd. Why? The probable reasons are many.

The first reason is that many do not like the attention. When you speak to a crowd, all eyes are on you. If you mess up, everyone sees the mistake. If you go blank and do not know what to say, everyone sees. Those thoughts make many people uncomfortable, and they try their best to avoid such a situation.

137

The second reason that people do not like to speak publicly is that they think they have nothing of importance to say. Being unprepared is a key reason that people who are asked to speak to a crowd have nothing to say.

This chapter gives tips on how to speak in front of a crowd (Figure 6-1). Although speaking does come naturally for some people, most would probably agree that public speaking is a learned skill. You too can learn how to talk in front of crowds no matter what subject is thrown your way. The key to public speaking is practice.

You have heard the phrase used before: Practice makes perfect. With public speaking, nothing is more truthful. In order to become an effective speaker, you have to work at it. Constant practice, delivery, and critique are what effective speakers engage in continually.

Pay close attention to the lessons in this chapter. They will guide you in becoming more articulate, more convincing, and more effective as an agricultural communicator.

FIGURE 6-1 This chapter teaches you how to speak in front of a crowd. (*Courtesy of Getty Images/PhotoDisc*)

MAKING THE MOST OF YOUR TIME IN THE PUBLIC EYE

As a communications professional in the agricultural industry, you may be called on to speak in front of trade groups, to customers, or to the public at large. You have a responsibility to make the most of the time you have been granted in the public eye.

For companies, celebrities, and politicians, the public eye can do one of two things: break careers or boost them (Figure 6-2). As a professional, you always want to put yourself in the position to boost your career as well as advance the goals of the company you work for.

Publicity can be a double-edged sword. On one hand, you would like your company to be featured in positive news stories because any mention of it or its products by news outlets is free advertising. On the other hand, you never want the kind of negative publicity that can put companies out of business or hamper their earnings.

A few years ago, talk show host Oprah Winfrey was taping a show about the mad cow disease epidemic in England. On her show, she had a public health advocate who stated that if mad cow disease made it to the United States, the results would be so devastating it would make the AIDS virus look like the common cold. After that statement, Winfrey said that she would never eat a hamburger again.

FIGURE 6-2 Politicians know that the public eye can help them achieve their goals. (*Courtesy National Archives*)

Since Winfrey is such a beloved public figure, a panic was set off in the financial institutions where cattle futures are traded. As a result, the cattle industry suffered extreme financial losses until the public became convinced that the nation's meat supply was safe. Negative publicity can hurt. As a speaker, it is your job to make sure that accurate facts about the people you represent are brought before the public.

Therefore, the trick to making the most of your time in the public eye is not a trick at all. It is a matter of providing relevant and substantive commentary on whatever subject you represent. Think about it this way: if someone were to ask you about yourself, would you spin the truth or provide facts? As an agricultural communicator, you have the same responsibility.

THE ROLE OF THE PUBLIC SPEAKER

If you work for a biotechnology company that is conducting research on a forage plant that will yield 5 percent higher protein than conventional forage plants, you certainly would want the world to know about it. What would you say? Moreover, what would you say if an agricultural reporter called and wanted the complete details on how the plant would change the cattle grazing industry?

As an agricultural communications professional, you have the duty to be able to answer such questions should they arise. You also have the unfortunate obligation to respond to bad news should the company you work for delegate that responsibility to you.

Ultimately, the public speaker has many roles within the realm of agricultural communications: motivator, salesperson, **spokesperson**, spin doctor, and damage control specialist. To wear all of those hats, public speakers need several tools in order to be effective (Figure 6-3).

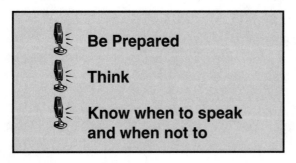

FIGURE 6-3 The tools of the public speaker.

- **Be prepared.** Preparedness is the best tool to have in your repertoire. You have the duty to stay abreast of all company news and developments so that you can be thorough and give educated answers to any questions posed to you.
- **Think.** The ability to think is a powerful tool. Using your head and thinking answers through before delivering them can ultimately save you and your company lots of headaches. When asked a question, take a moment to think; then speak. Being prepared goes a long way to helping your cognitive processes.

Being a public speaker requires the ability to think quickly.

- **Know when to speak and when not to.** If you watch television shows about the criminal court system, you will invariably notice that a defendant will refuse to answer questions because his answers many incriminate him. The same applies to your role as a public speaker. Just because you know all there is to know about your company does not mean that you *should* tell all there is to know. Although it is hard to speak in generalities on this point, as a speaker you will have to find out where the lines lie on what you can say and what you cannot.

USING YOUR TIME WISELY

When you are given the charge to speak, use that time as effectively as possible. Like publicity, time can be a double-edged sword. If you have fifteen minutes to give a speech and you use all of that time, did you say everything you needed to say? Alternatively, did you use the fifteen minutes because you had it yet gave only four minutes of relevant information and rambled the rest of the time?

Time is a precious commodity. Use it wisely.

Making the most of your time as a public speaker means saying what you have to say in a flowing, coherent manner. If you have fifteen minutes to speak and can deliver a solid message without any fluff in ten minutes, do so. Do not use the extra time just because you have it. Learn to be your own editor, and end when you have said everything you set out to.

If you have fifteen minutes to speak and twenty minutes of material, you must find a way to cut five minutes off what you have to say. By incorporating the inverted pyramid style of writing speeches as mentioned in Chapter 4, shortening your presentations will be easy.

View your time to speak in public as a precious commodity. It is your chance to highlight what you are promoting and get the facts out about a topic from your or your company's point of view.

SPEAKING EXTEMPORANEOUSLY: A FEW TIPS

You may be asked to get up and say a few words without any prior knowledge of the invitation or topic (Figure 6-4). This type of speaking is called

FIGURE 6-4   You may be called on to make an impromptu speech. (*Courtesy USDA/ARS*)

extemporaneous or impromptu. **Extemporaneous speaking** is defined as a type of speech in which the speaker has only a few moments to prepare. An impromptu speech is given spontaneously without any preparation time. Often, these two speech types are perceived to be the most difficult of all of the speaking tasks since speaking extemporaneously or impromptu requires you to think on your feet quickly and thoroughly.

You have seen how this happens, no doubt. Someone wins an award, goes to the podium to accept it, and the crowd starts to yell, "Speech! Speech!" Before you know it, the person on stage is rambling on, thanking the people who helped him or her win the award. Often, these types of speeches start and end abruptly, with little, if any, continuity in the delivery of the message. Shaky delivery in an extemporaneous speech goes back to the point mentioned at the start of this chapter: people's general fear of the spotlight and being the center of attention.

As an agricultural communicator, there is no reason for you to fear speaking publicly. Although you may not know much about the topic, there are techniques you can use to help you prepare and deliver an impromptu message well before you are even asked to step up to a microphone.

If you become an FFA member, you will have the opportunity to speak extemporaneously in the National FFA Extemporaneously Speaking Program. The extemporaneous speaking program is an event that allows members who participate the chance to use their speaking, organizational, and thinking skills in one single contest. Being in the extemporaneous speaking contest involves picking a random topic at contest time, preparing a speech during a 30-minute time slot, and delivering a four to six minute speech to a panel of judges. For more information on the FFA's extemporaneous speaking contest, check out <www.ffa.org>.

Doing Research

Successful extemporaneous speeches hinge on quality research. When communicating for the agricultural industry, it is in your best interest to be prepared to speak at all times. That means that you should keep up to date on all pertinent information concerning the sector of the industry you are working in. For example, many agricultural science teachers are asked to talk about their FFA programs spontaneously. By understanding the intricacies of their local program, as well as a vast knowledge of state and national FFA policies and procedures, a teacher has all of the background material needed to deliver a five-minute or longer address on the topic.

If you are an extemporaneous speech competitor, you must do research ahead of time because you cannot always anticipate the topic accurately. One of the best techniques to use to prepare as an extemporaneous speaking contestant is to look for topics that are current in nature such as the agriculturist's role in providing environmental protection and thoroughly research them. This topic is timeless.

You must also do your research quickly. In the National FFA Extemporaneous Speech Contest, you have only thirty minutes to research, craft, and deliver a speech. Therefore, your time spent researching is precious, and it is in your best interest to know your material ahead of time.

Making an Outline

Good extemporaneous speakers begin with a solid outline, prepared mentally or written on paper. An outline helps you organize your thoughts in a flowing and coherent manner. It also provides a framework for planning speech length. Length of speech is especially important in the National FFA Extemporaneous Speaking Contest since the rules limit all speeches to four to six minutes.

For an extemporaneous speech, there is a basic format for organizing your thoughts that is used universally by most extemporaneous speakers (Figure 6-5).

- **Step 1: Narrow the topic.** After you have selected a topic (as in the case of a contest) or been given one by, say, the chairperson of a civic club, the first step is to narrow it. If you were asked to speak about the local FFA chapter, for example, what would you say? Many chapters are involved in lots of activities, and someone from those chapters could talk for an hour or more about those activities. Therefore, you narrow the topic and then develop it.
- **Step 2: Come up with an attention getter.** All speeches start well with an attention getter to focus the attention of the audience. Attention getters come in many forms: a story that relates to the topic of the speech, a thought question that challenges the audience to contemplate the subject, or a prop, such as an FFA jacket, that prompts curiosity and discussion, for example. By captivating the audience from the first sentence, you can keep them hanging on every word you speak. The key is to get their attention.
- **Step 3: Think of three to five points.** Once you have narrowed the speech and developed the attention getter, it is time to come up with points you want to make about the topic. Points in a speech give context and meaning to a speech or conversation and prevent it from being a rambling monologue. Each point in the speech should relate to the topic at large and act as a mini-speech.

  Before expanding on each point, tell the audience the points you are about to make so they know what is coming. Three to five points

Extemp Outline

Topic: How the FFA benefits youth

Introduction: Introduce yourself and the topic

Point 1: FFA benefits youth by giving then a chance to associate with like-minded people

Point 2: FFA benefits youth by preparing then for careers in agriculture

Point 3: FFA benefits youth by helping them secure college scholarships

Recap all points

Conclusion: Thank everyone for their time and ask for questions from the floor

FIGURE 6-5 Making an extemporaneous speaking outline is an easy process.

make a presentation multidimensional. In addition, having more than one point lengthens the speech because it is always easier to say a few relevant words about a single point than to expound on a single point.

- **Step 4: Conclude the speech.** Finally, wrap up all of the points you made and conclude in a logical manner. When concluding, you might want to allude back to the story you told at the beginning of the speech and relate it to the points you made. This method ties the whole speech together and gives the message depth and relevance.

Here is the format for creating extemporaneous speeches.

Title of Speech _____
Introduction   _____
Point 1   _____
Point 2   _____
Point 3   _____
Conclusion   _____

This basic template can apply to just about any extemporaneous speech you may give. In the case of the FFA speech, you can simply fill in the blanks.

Title of Speech: The Anytown FFA—Speaking Activities

Introduction: Tell story about a girl who used to be afraid to speak in public.

Point 1:   The Prepared Public Speaking Career Development Event (CDE)

Point 2:   The Extemporaneous Public Speaking CDE

Point 3:   The Junior Prepared Public Speaking CDE

Point 4:   The Marketing Plan CDE

Point 5:   The FFA Creed Speaking CDE

Conclusion: Relate the story you told at the first of the speech to the person standing giving the speech.

If you have the challenge of speaking for five minutes and use this model, you will speak on each part of the speech for 43 seconds. Looking at it that way makes the challenging task of speaking extemporaneously seem much easier.

Delivering the Speech

When I was in college, one of my professors told me that the way to address a group was a simple 1-2-3 process.

1. Tell them what you are going to tell them.
2. Tell them.
3. Tell them what you told them.

This was his clever way of outlining the model for teaching. In delivering any kind of address, state your points, elaborate on the points, and then recap the points. In that way, your message is hammered home to your audience in a subtle yet effective way.

When using this technique, it is important to make your points clear. You might want to say something like, "In this speech, I'd like to tell you five reasons that biotechnology is good for agriculture." You can end the speech by saying, "As I mentioned in my presentation, biotechnology is good for agriculture because [state the five points]."

Physical delivery of a speech is also important. When giving a speech, extemporaneous or otherwise, you always want to be relaxed (Figure 6-6). If you are relaxed, your audience will also be relaxed and ultimately

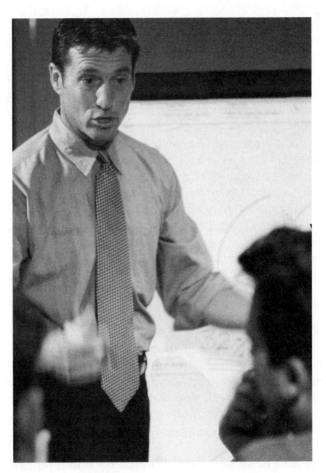

FIGURE 6-6   When giving a speech, be relaxed.
(*Courtesy of Getty Images/PhotoDisc*)

more attentive. Give your speech in a relaxed manner using natural gestures. Avoid overacting, which is an unnatural and unconvincing style of delivery.

SPEAKING FROM PREPARED NOTES

Speaking from prepared notes, like extemporaneous speaking, is an acquired skill. The best speakers today were not born knowing how to speak publicly. They practiced their craft in front of crowds and perfected the art of delivering a powerful message to anywhere from a few people to thousands.

Speaking from prepared notes is often called **prepared public speaking** (Figure 6-7). In the National FFA Organization, prepared public speaking is a competitive event. The rules for this contest state that the speech can be on any agricultural topic and should run six to eight minutes in length.

The best speakers always have one thing in common: a dynamic way of presenting information. Like a salesperson, they use voice inflection and emotion to sell the points of their speech in a way that draws the audience in with every word. A good speaker also can capture the audience's imagination by telling funny or heart-wrenching stories. Stories, analogies, and dynamic speech patterns are all ingredients in a powerful presentation.

Good speakers are not born; they are made. To be a good speaker, you have to practice. Practice often involves many hours of writing, rewriting, and rehearsing, and, depending on the nature of the speech, memorizing. Nevertheless, a good speech starts with a solid concept and topic. There are some techniques you can learn for developing solid outlines and writing convincing speeches.

Working Up an Outline

Like speaking extemporaneously, writing a speech from prepared notes has an unwritten protocol that nearly all successful speakers follow. Take a moment sometime to surf through your television channels and watch televangelists, political candidates, and other speakers whose sole purpose is to deliver their message. If you watch long enough, you will see patterns in their speeches.

When working up an outline for a speech, keep these patterns in mind.

Start with a joke or some other icebreaker   When you stand up to speak after being introduced, the audience is waiting to hear what you have to say. Just as you seem to have butterflies in your stomach, some of the

FIGURE 6-7   Prepared public speaking involves talking from notes. (*Courtesy Russell A. Graves*)

audience may feel the same way too. They want to hear a good message but are not sure what to expect.

As a speaker, your job is to relax the audience. They want to feel as if you are speaking to them, rather like from one friend to another. You need to break the ice. An icebreaker may be a fun activity that gets people laughing or relaxed, or it could be a joke or a story that is appropriate for everyone in the audience. The key is to get the audience relaxed so that they are in tune with what you have to say.

As a speaker, one of your roles is an icebreaker for crowds.

Segue into your speech   Once you have spent a minute or two on the ice-breaker, it is time for a segue into the heart of your speech. The segue is a sentence or two that redirects the audience's attention from the ice-breaker to the next phase of the speech. Sometimes it is a story that relates to the points you are about to deliver or an abrupt change in the tone of the speech. Find out what works best for you, and develop a segue that adds to the impact of your speech.

Outline the points   Tell the audience what you plan elaborating on in your speech. Instead of dropping your message on them, prepare them for what they are about to hear by giving a brief outline of the main points you will make. In this way, you prepare them for the next phase of the speech so they are ready to listen more attentively or take notes if they choose.

Explain each point   Once you have outlined the points, start the process of crafting your argument on each one. Do not just set out each point and then move on. Take the time to explain your point of view. This explanation will help build the audience's confidence in your credibility as a speaker.

With each point, explain your stance by using analogies, stories, or jokes. Again, one of your jobs is to keep the audience awake and paying attention. As a speaker, you are just as much an entertainer as you are a speaker, so keep that point in mind by adding color and flair to the entire speech.

Recap the points   Before you start the closing portion of your speech, keep in mind that you should recap all of the points that you just made. The reason for the recap is simple: to help the audience remember what you said. You no doubt can recall teachers who used this technique, which they probably referred to as a review.

A recap does not have to be lengthy. A minute or so is fine. Just make sure the audience knows the points you just addressed and has an understanding of them.

End with a bang   Your goal as a speaker is to have people remember you and your message and be invited to more speaking events, so when you end,

end strong. Finish with a funny story, a song quotation, or a reflective story. When you are done, thank everyone for their time and wish them a good day.

Writing a Speech

For many speakers, perhaps the hardest part of the process is to pen a thoughtful and meaningful speech. (The U.S. president employs a battery of speechwriters to prepare his addresses.) The process of writing a lengthy essay on a topic on which you will speak can be a daunting one. Nevertheless, there are some techniques you can employ to make the process smoother.

After you have developed an outline, address each point mentally first, and try to come up with something meaningful to say on it. A mental idea of what you plan to say goes far in helping you write the first draft of a speech. As you begin to write, you can refer back to those mental, as well as written, notes to help the process along.

When you write the first draft of a speech, write as you talk, with incomplete sentences and all. That is not to say that you should accept bad grammar. Instead, it is just a technique to employ while working on the first draft. The bad grammar can be fixed later in the editing process. By transferring what you are thinking on to paper, you can write much more efficiently than if you try to write in a formal manner the first time.

Most writers recognize that they have two voices in which to communicate: the voice that they speak to people every day with and the voice they write with. The former is a less formal way of speaking with incomplete sentences and figures of speech and colloquialisms that ordinarily have no place in formal writing. Their written work is much more precise and formal. The reason is that it is harder to communicate with the written word.

With the spoken word, voice inflection and hand gestures help get your point across. With writing, you have no such luxury. You must use a bland canvas—a piece of white paper with black type—to convey emotions and sincerity. Humor, for example does not come across as funny when it is read as it does when spoken. The speaker's stage presence has a lot to do with getting the crowd in the mood to laugh at jokes she may tell. Give the same audience a written transcript of the jokes, and chances are that most will not think they are as funny.

Following is a speech that was written for the FFA Prepared Public Speaking Contest. This is a draft that was revised into its final form before the speaker presented it at the contest.

The speech works well in tying an emotional element to the importance of FFA in public schools. Filled with facts, the speech also puts a personal spin on what the FFA means to the speaker. As you read the speech, pay close attention to the way the writer ties in facts with emotional elements.

A Girl . . . A Dirt Road . . . and the FFA

(Courtesy of Mary Katherine Henderson)

I've done it so many times . . . I could do it in my sleep. I walk out my back door, jump in my old Chevy truck, and start it up. I then put it in first and head north down a long dirt road for the 20-minute drive to school. I don't really mind the drive because I always have something to think about, and it's usually something to do with the FFA. Every day this drive has the same three things in common: a girl, a dirt road, and the FFA.

You see for the past three years, the FFA has been a big part of my life. I joined as a shy, quiet freshman who couldn't stand in front of my agriculture class to say the FFA creed. Over these last few years I have been involved in many activities, from meats judging to Senior Quiz to agricultural communications. The FFA has helped me to build my self-confidence and enjoy many new experiences.

Unfortunately, many students may not have the same opportunities that I've had, many schools are trying to close down their FFA programs and do away with agricultural education. In the budget proposal made by the governor of California in January of this year, there was a statement that suggested the removal of all vocational education programs from the state's high schools. This would severely hinder the national organization, with a loss of almost 50,000 members. This proposition quickly died as over 15,000 calls, letters, and e-mails were received by the governor's office within two days of this announcement.

Many people will hear stories such as this and think that its just terrible, but they won't understand the full impact of these attempts until it happens closer to home. Sadly it has. In order to meet budget cuts for the next school year both the Groom and Texline Independent School Districts were compelled to close down several of their programs, including agricultural education. This means the students at these schools will no longer have the opportunity to participate in this great organization.

This brings me to the point of my speech: Is FFA and agricultural education important in today's society?

Purdue University conducted a study in 1999 that compared the attitudes, behaviors, and experiences of "typical" high school students to students involved in the FFA. Comparisons were made from a study, "The State of Our

Nation's Youth" conducted by the Horatio Alger Association. This study provided proof of the positive impact the FFA has on teenagers.

These studies showed that 90 percent of all FFA members are active in extracurricular school and community activities compared to only 77 percent of typical non-member students. Only 73 percent of typical students feel that it is important to do their best in everything, and a mere 66 percent relate personal effort to personal success. Comparatively 93 percent of FFA members feel it is important to do their best at everything and 91% relate effort to success.

Education is very important to the future of our nation but only 32 percent of typical students think their courses are exciting versus 83 percent of students in the FFA. This study has also found that 86 percent of FFA members plan to go to college, while only 84 percent of typical students do.

Aside from all of these facts is the true heart of the study. Larry Case, the national FFA advisor, says "Anyone who has been around FFA members know these students are positive and energetic about their education and their future." In short, FFA members feel good about themselves.

This once again brings me to the question: Is the FFA important to today's society? I believe that it is.

In the case of California this would have been a tragedy for two reasons. One, California is the largest producer of agricultural products in the nation. Carrie Robertson, an agricultural science teacher at a Los Angeles school, put it "Every other subject exists because of agriculture; people need to understand that the strength of any society is its ability to feed itself." And secondly, because California has the second largest state membership in the nation, with over 48,000 members, second only to Texas by 8700 members. Where would all these students be without the FFA? Whose to say. If the largest agriculture producing state in our nation doesn't have the people with the knowledge and skills to feed us, then where does that leave us?

As for schools such as Groom and Texline, matters don't look quite so bright. These towns don't have much of a choice, cut programs or close the school and force them to travel anywhere from 10 to 40 miles to school everyday. This is truly a tragedy for these schools because the FFA has so many possibilities for small town students that would otherwise be unattainable. Opportunities such as those to travel all over this great state for different contests and activities, and the opportunity to receive large amounts of scholarship dollars that under any other circumstances would not be available to them. As the comparison between FFA members and non-members proved, students in these schools are left without much excitement or hope for their futures because they have nothing to look forward to every day.

As for this girl? Well I would be devastated if the FFA were ever removed from my school. What would I have to look forward to at the end of my long drive to school? And what about all those other students, across the state and nation, who have long dirt roads to travel every day? Or even those who live just around the corner? Where would we be without the FFA? I believe that I would still be the same shy quiet girl I was as a freshman, who was too afraid to even stand in front of a group of her peers in order to recite the FFA creed. And I certainly wouldn't have the courage to be giving this speech to a few people I don't know. Yes, I'm still only a girl, but I am so much more because of a dirt road, and the FFA.

As you read the preceding speech, you might have noticed that it flows nicely from one point to the next. Unlike a feature story, it does not start off with a definite lead and is not written in inverted pyramid style. Instead, the speech meanders through the points and lets the audience ponder what is coming next—an effective speech technique.

It used another good technique too: starting with a story. The story sets the tone for the speech and helps emphasize the main point of the speech, which is the importance of FFA in the public schools of America.

After the first half of the speech, the thesis of the address is backed up by facts about California and cites a study by an organization that emphasizes the importance of the FFA. In the end, the speech is tied back together by citing, once again, the story that led off the speech.

Speeches such as this one try to take the audience on an emotional roller-coaster ride. From happy and jovial to serious and reflective, this speech has many of the same elements that the nation's top speakers employ. Ultimately, the main goal of any speech is to leave an impression with the listeners—an impression that stays with them and keeps them talking about you long after your speaking engagement is over. If you can master that one aspect, you are well on your way to become a master at oration.

Rehearsing

You have heard the saying, "Practice makes perfect." Most often, this phrase is associated with athletic events, and rightly so. It also applies in the realm of speaking as a solid reminder of the responsibility that people who are hired as a speaker hold.

Before giving any prepared speech, make sure you practice. Practice can take many forms and varies from person to person. For starters, keep these points in mind.

Read the script   When you start your practice sessions, begin by reading the script silently and then aloud (Figure 6-8). Reading the script repeat-

FIGURE 6-8   When you start to practice your speech, read the script thoroughly. (*Courtesy Russell A. Graves*)

edly will help ingrain the material into your mind. It will also help you discover patterns in the speech, like natural breaks or changes in the subject, and will aid in the memorization. When you first begin to read the speech, it does not matter whether you put any emphasis into the delivery. The key here is to commit the content to memory.

Memorize   For many, this second rehearsal step is the toughest. Memorization takes time because it is difficult to memorize a thousand or more words of text. Unless you are blessed with a photographic memory, the best technique for memorizing something is to read and recite it repeatedly. If you have good ad-libbing skills, you may want to memorize the speech's outline and ad-lib through each point. A word of caution, though: only experienced orators should use this technique. Imagine how embarrassed you would be if you tried to ad-lib a speech and instead stood there with nothing to say.

Practice by yourself   Once you have memorized the speech, turn to putting the presentation together. Do this by yourself. Think about how to emphasize different words and phrases so that you can get the most impact from them. Say your speech repeatedly in your room, in your backyard, or anywhere else that you can be comfortable and away from distractions.

At first, you may feel goofy about speaking aloud with no one else around, but you must remember that this step in preparing a speech for a public audience is perhaps the most important. Good speeches do not just happen. Remember that practice makes perfect.

Practice in front of a friend   Once you feel comfortable with your delivery of the speech, it is time to perform for an audience. How big an audience is up to you—it can be one person or several. The key is to choose people who are not shy about giving you critical feedback on your performance. Have your audience judge you on how well you articulate each word, your eye contact, your body and hand movements, and the speed at which you speak. The purpose of practicing in front of an audience is to work the bugs out of your presentation in front of a few people instead of messing up when there is an auditorium full of attentive listeners.

Use visualizations   When you are not actively practicing the speech by saying it aloud, visualize yourself in front of an audience giving the speech. If you watch the skiing events in the Winter Olympics, you have probably noticed skiers with their eyes closed and moving their hands in rhythmic fashion as they visualize themselves running the downhill course. The thinking behind this technique is that if they picture themselves running the course, it will be virtually automatic for them when they actually start the race. The same technique applies to speaking. Visualize yourself in front of an attentive audience giving the speech of your life. The practice will pay off when you feel as if you have given the speech a hundred times before.

The final delivery   After all the rehearsal, it eventually comes time to deliver the speech. The operative word at this point is *relax*. No one has ever died from the nervousness associated with giving a speech. If you have taken the time to practice and be critiqued by a set of fair-minded friends, then your speech ought to be automatic. If you take the time to prepare the best you possibly can for a speech, you indeed can leave a positive mark on your audience.

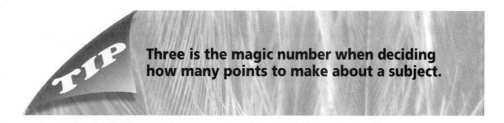

Three is the magic number when deciding how many points to make about a subject.

WORKING AS A SPOKESPERSON

Sometimes as an agricultural communicator and because of your ability to deliver and articulate facts, you may be asked to serve as a spokesperson for your FFA chapter, your school's student council, or an employer. When asked to serve as a spokesperson, do not take this responsibility lightly. It is a very important job because you represent something much bigger than you are.

The public's eyes are on you as a spokesperson. Therefore, it is your responsibility to deliver accurate and thoughtful responses to questions asked by reporters or the public at large, all the while, delivering your message in an honest and professional manner.

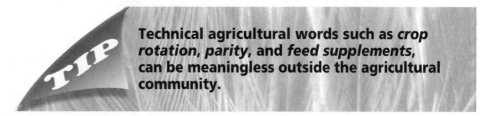

Technical agricultural words such as *crop rotation*, *parity*, and *feed supplements*, can be meaningless outside the agricultural community.

When things are going smoothly for an organization, the spokesperson's job can be enjoyable and easy. When things get tough, it is a different story. If all the news you had to communicate was positive, then everyone could be a spokesperson. Sometimes, however, you may have the job of delivering bad news. This situation calls for a person who is honest under pressure, calm, forthright, and dependable.

Although your job title may not say it, you will be the spokesperson in many jobs that you may take in the agricultural communications field. A prime example is that of agricultural science teacher. As a teacher in this field, you know that communication is a daily chore. You have to deliver your lessons every day to your classes. Your role as a spokesperson comes in when you speak about the FFA to the school's faculty and administration.

In a way, we are all spokespeople for the most important entity there is: ourselves. Your role as a spokesperson is an extension of what all of us

FIGURE 6-9   As a spokesperson, you should project a positive image. (*Courtesy USDA/ARS*)

should spend time doing daily. Thinking of ourselves and the people around us in a positive light gives other people confidence in us. Think positively, and others around you will do the same (Figure 6-9).

HANDLING QUESTIONS FROM THE FLOOR

After you deliver a presentation, whether it is a press conference or a prepared public speech for an FFA competition, you will often have to open the floor for questioning (Figure 6-10). Sometimes the questions will be easy and kind in nature, as in the case of someone allowing you to expand on a point you already made.

TIP   **Always think before you talk.**

FIGURE 6-10   As a speaker, you are often asked to handle questions from the floor. (*Courtesy USDA/ARS*)

Other times, the questions may be tougher. A questioner may want to know why your company chose to use the services of a company that is under investigation by law enforcement officials. How would you answer a tough question like that? Many people may be tempted to say "no comment," but that may not always be the correct answer to give because it may plant a seed of doubt and decisiveness in the mind of your audience.

The basic rule of a spokesperson is to know what to say and when to say it. Unfortunately, there is no formula for knowing how much information you should divulge. This should be handled on a case-by-case basis.

Think Before Talking

Perhaps one of the biggest mistakes people make is answering a question without thinking of the possible ramifications that answer may have. Consider many politicians. When they have given less-than-truthful answers on a subject and then later the truth becomes known, they get into the proverbial hot water. Chances are that when someone posed the tough question, they responded with an easy answer before they thought of the ramifications their response would bring.

When answering questions, always be truthful. Take a moment to think of what you want to say before you answer. However, do not stand there quietly as everyone stares at you, which can raise the tension in the room. Instead, buy yourself a little time; ask for the question to be repeated or ask for a clarification on the question. You may also want to clarify the question for yourself by stating, "So what you are asking is . . ." These simple techniques can give you time to think of a response before giving an answer that will go on the record for perpetuity.

Never answer questions in a way that leaves the door open for you to be questioned on an area that you are not prepared to give an intelligent answer on. Everything is attributable. When quoting figures, it is preferable to start the sentence with, "According to [cite your source] . . ." By your attributing, the audience gets the sense that you are passing along someone else's information. You never want to get into the habit of having to answer for someone else's research.

Always attribute information.

The best technique you can use when faced with the challenge of answering questions is to be prepared. If you are giving a prepared public speech for a competition and have studied all aspects relating to the subject, then you should be prepared to give an effective answer on any related question. If you studied your material poorly, then your ability to answer questions is limited. The bottom line is to be prepared and rehearse answers before you have to give them. Above all, think about what you will say.

Delivering an Effective Answer

You have thought about your answer; now it is time to deliver a response. The best answers are ones that repeat the question. That way, the people who may not have heard the full question will understand the line of questioning presented. Here is an example.

Reporter: "Mr. Jones, you spoke about your FFA chapter and said that you will be competing in six CDEs this spring. Could you tell me what they are?

Jones: Thanks for the question. The six CDEs we will be competing in this spring are livestock evaluation, horse evaluation, marketing plan, prepared public speaking, agriculture issues forum, and the job interview competition. Are there any other questions?

Jones started his response by thanking the reporter for the question. He did that to buy a little time to think. When Jones answered the question, he did so in a straightforward manner and did not leave himself open to be trapped by a subsequent question that he may not be ready to answer.

Effective responses are the best tool in leaving the question-asking audience with a good opinion of you and the entity you represent. Take the time to practice answering techniques so that when the time comes, you will be prepared.

THE ILLUSTRATED TALK: HOW TO ADD VISUALS TO PRESENTATIONS

Well-thought-out and professional visual aids can do more to enhance your speech than perhaps any other technique. A good photograph or a well-planned illustration can communicate a message in an instant.

Visuals appeal to the learning styles of many different people and broaden your audience. Some people learn by listening, some by reading, and others learn by tactile means. (Tactile learners must touch and handle objects related to the subject in order to get a grasp of the concepts. Tactile learning can be described as the learning-by-doing approach.) People who learn by hearing can listen to you, while people who are visual learners can cue in on your visual presentation while listening to you (Figure 6-11). Tactile learners can be presented with materials if the presentation warrants.

Visuals also give the audience a chance to focus on something besides you. If the presentation is long, people may become bored by listening to you speak after a time, no matter how dynamic you are. Visuals help break up the flow (Figure 6-12).

Visuals also help solidify points that you may be speaking about. For example, if your speech were about the effects of weather on cotton yields, a visual showing the annual rainfall amounts would be appropriate.

FIGURE 6-11   Some people are visual learners and can gather a lot of information from one picture. (*Courtesy Russell A. Graves*)

Visual aids come in many forms. They can be a laptop computer connected to a projector showing a PowerPoint presentation; a slide show outlining the various breeds of beef cattle; posters; flyers; or anything else you can think of. Be creative.

A laptop computer can be connected to a projector and be used as a visual aid.

Introduction

Photography can be used for:

- Agriscience Fairs

- Ag Communications class

- Ag Issue presentations

- FFA Skills presentations

- Recording CDE contests — classes of livestock, cuts of meat, plant ID, etc.

FIGURE 6-12   Visuals help keep the audience's attention.

TIP   **Slide projectors are effective ways to present visual information.**

As powerful as visuals can be, they should not be a crutch to help you get through a presentation. Visuals are meant to enhance your verbal presentation, not be the sole focus. Many beginning speakers rely too much on visuals to the point that they become distracting. A good rule of thumb is to have just a few visuals to illustrate the major points of your speech and not the minor ones.

Above all, practice. Adding visuals to a presentation increases the complexity. You have to remember when to change the slide in the projector or when to reveal the next flip chart, so it is crucial to work on your timing beforehand. Poorly executed visuals can only lead to your audience's remembering your mistakes.

CONCLUSION

This chapter has provided ways to make you an effective oral communicator. It does not matter if you were born shy or outgoing; everyone can be a good speaker.

This work begins with having the knowledge it takes to develop good, solid outlines, write crisp text, and deliver a meaningful address.

When making an outline, first narrow the topic down to a detailed subject. Then, instead of looking at it from a broad standpoint, think of the topic in terms of more narrowly focused points. With every subject, there is a story to tell. That story may not be immediately apparent, but with research, story lines begin to develop.

After you have developed a solid outline, fill in the blanks with good conversational writing that helps you express yourself. It does not matter so much if the first draft has some grammatical errors; you can fix those later.

Once the speech is written, rehearse it. Practicing a speech builds confidence. Once you know the material, you can deliver it at a moment's notice. Practicing requires repetitive recitation of the speech, but your audience deserves your best. Complete and concentrated practice is necessary.

Sometimes after giving a speech, you may be asked questions. Never answer questions in a way that will trap you. Give direct, forthright answers free from speculation and attributable to direct sources. Never allow conjecture to seep into your responses, and always, always stick to the facts.

As a spokesperson, your job is to represent yourself and the people you represent in the most respectable manner possible. Always be a professional when representing yourself. Professionalism shows, and so does confidence. Confidence in yourself is contagious. If your audience knows that you believe in yourself, they will believe what you have to say as well. Good luck—and remember that no one ever died from nervousness.

REVIEW QUESTIONS

Multiple Choice

1. Publicity can be
   a. hard to produce.
   b. easy to obtain.
   c. a double-edged sword.
   d. a great way to increase revenue.

2. The best way to make the most of your time as a public speaker is to
    a. get to know your subject.
    b. stay relaxed.
    c. provide relevant and substantive commentary.
    d. all of the above

3. The public speaker has many roles; some of them are:
    a. salesperson, spin doctor, and damage control expert.
    b. reporter, expert in a particular field, and upper-level management.
    c. jack of all trades, part-time psychologist, and quick thinker.
    d. none of the above

4. Part of making the most of your time as a public speaker means
    a. delivering your speech in a flowing and coherent manner.
    b. being able to come up with answers on the fly.
    c. knowing everything there is to know about your subject.
    d. being able to relate to the common people.

5. Extemporaneous speaking is defined as
    a. a type of speech that one has a few minutes in which to prepare.
    b. giving a spontaneous speech without any preparation.
    c. being able to speak on a wide range of topics.
    d. none of the above

6. An impromptu speech is defined as
    a. a type of speech that one has a few minutes in which to prepare.
    b. giving a spontaneous speech without any preparation.
    c. being able to speak on a wide range of topics.
    d. none of the above

7. One of the best techniques to use in extemporaneous speaking is
    a. researching topics that are stale or outdated.
    b. getting as much information about as many topics as possible.
    c. researching topics that are current and thoroughly researching those that are timeless in nature.
    d. none of the above

8. Good extemporaneous speakers always begin with a(n)
    a. good outline.
    b. good joke.
    c. anecdote that pertains to the subject.
    d. all of the above

9. The use of voice inflection and emotion can

   a. make people connect with your subject on an emotional level.
   b. help sell the points of a speech.
   c. keep people guessing what the main point is.
   d. none of the above

10. An icebreaker is

    a. a ship usually stationed in the arctic.
    b. useful when the pipes freeze.
    c. a fun activity that gets people laughing and relaxed.
    d. all of the above

11. A segue is a

    a. story that can lead into any topic.
    b. short blurb that explains the headline of the speech.
    c. story that relates to the points you are about to deliver.
    d. none of the above

12. During a speech, the main points should be

    a. given and then you should move on to another topic.
    b. given at the beginning and then elaborated on.
    c. explained to show your point of view.
    d. all of the above

13. Before closing a speech,

    a. thank the audience for their attention.
    b. give the location of your next appearance.
    c. recap all of the points that you made during the speech.
    d. remember to be gracious for the applause that you are about to receive.

14. One of your goals as a speaker is to have people remember you and your message. One way to attain this goal is to

    a. be forceful in your delivery.
    b. finish strong.
    c. be as funny as possible during your delivery.
    d. have at least eight points to address during the course of your speech.

15. By transferring what you are thinking onto paper, you

    a. can write much more efficiently.
    b. will be able to write a speech that will sound more like an everyday discussion.

    c. will be able to memorize your speech more efficiently.
    d. none of the above

16. The best technique to memorize something is to

    a. write it by hand, over and over.
    b. alternate writing it and reading it several times.
    c. read it and recite it repeatedly.
    d. all of the above

17. Good language arts are

    a. fairly important for the spokesperson.
    b. a good tool to have in one's arsenal as a spokesperson.
    c. helpful to the spokesperson.
    d. essential for the spokesperson.

18. As a spokesperson, one must

    a. know a little about all facets of the organization.
    b. be able to recite the job descriptions about each employee of an organization.
    c. be well informed about the total organization.
    d. none of the above

19. A basic rule for a spokesperson is

    a. to know what to say and when to say it.
    b. have notes on the subject he or she is about to speak on.
    c. to be courteous and forthcoming.
    d. not to divulge any information that is not necessary or that can hurt the company that he or she represents.

20. Visuals appeal to

    a. most people.
    b. the learning styles of many different people and broaden your audience.
    c. everyone.
    d. should never be used.

21. Visuals are powerful and

    a. should be used as much as possible to help you get through your presentation.
    b. can help get your point across.
    c. should never be used as a crutch on which you lean on to help you get through a presentation.
    d. none of the above

Fill in the Blank

answering	icebreaker	public
attention	worth	questions
beforehand	delivery	redirects
comfortable	answering	repeatedly
consideration	argument	representation
crafting	conveys	solidify
defamatory	relaxed	analogies
extemporaneous	reading	jokes
first draft	responses	outline
frame	otherwise	express
framework	deserve	granted
hearing	technique	eye
impression	front	asked
interview	record	audience's
off	libel	ingrain
page	write	unjustly
picture	elaborating	speech
plan	visuals	stories
prepared	delivering	ultimately
preparedness	answers	words
questions	professional	

1. As a communications _____, you have the responsibility to make the most of the time you have been _____ in the _____ _____.

2. _____ is the best tool you can have in your repertoire.

3. Using your head and thinking answers through before actually _____ them can _____ save you and your company lots of headaches.

4. An _____ gives you a _____ in planning how long a speech will be.

5. When giving a speech, _____ or _____, you always want to be _____.

6. A segue is a sentence or two that _____ the _____ _____ from the _____ to the next phase of the speech.

7. Tell the audience what you _____ _____ on later in your speech.

8. Once you have told the audience the subjects you plan to speak on, start the process of _____ your _____ on each of the main topics the _____ possesses.

9. With each point, explain your stance by using _____, _____, or _____.

10. A mental _____ of what you plan to say goes far in helping you _____ the _____ of your speech.

11. _____ the script _____ will help _____ the material into your mind.

12. Once you feel _____ with your _____, it is time to perform for an audience.

13. By choosing _____ to _____ your thoughts _____, you will do a better job of communicating.

14. A spokesperson must convey a "feeling" that his or her ideas are _____ _____ and _____ serious _____.

15. Do not memorize _____, but think of _____ most likely to be _____ and _____ the _____ for them.

16. The absolute best _____ you can use when you are faced with the challenge of _____ _____ is to be _____.

17. _____ are important because they help _____ points that you may be speaking about.

ENHANCEMENT ACTIVITIES

1. Write topics dealing with agriculture on note cards. Then draw a card from the pile, and prepare an extemporaneous speech using the method prescribed in this chapter. First make a strong outline complete with points of discussion. Then deliver the speech, making sure it is four to six minutes long.

2. To prepare for Activity 1, compile a collection of current agricultural periodicals and books that will assist you in researching agricultural topics.

3. Come up with five possible attention getters for extemporaneous speeches.

4. Write and deliver a memorized prepared speech on a topic of your choice, taking note of the points offered in this chapter.

5. Practice your technique as a spokesperson by giving a short presentation to your class about a school club or activity you are involved in and then fielding questions about that activity from the floor.

6. Develop an illustrated talk using 35mm slides, flip charts, an overhead projector, video, or computer presentation software.

WEB LINKS

To find out more about the topics discussed in this chapter, visit these Web sites:

10 Tips for Successful Public Speaking
<http://www.toastmasters.org/>

FFA Extemporaneous and Prepared Speaking
<http://www.ffa.org/>

Impromptu or Extemporaneous Speaking
<http://www.ljlseminars.com/>

National Extemporaneous Speaking
<http://www.speechanddebate.com/>

Speechtips.com
<http://www.speechtips.com/>

Communication Skills from Iowa State University
<http://www.ag.iastate.edu/>
   (Search term: communication)

Tips for Creating Effective PowerPoint Presentations
<http://www.microsoft.com/>

Alternatively, do your own search at <http://www.alltheweb.com>
   (Search terms: public speaking, extemporaneous speaking,
      spokesperson, photography trends, agriculture communications)

BIBLIOGRAPHY

Instructional Materials Service. (2002). *Agriculture Communications 315 Course Materials*. College Station, TX: Author.

National FFA Organization. (2003). *National FFA Prepared Public Speaking Career Development Event*. Indianapolis: Author.

National FFA Organization. (2003). *National FFA Extemporaneous Speaking Career Development Event*. Indianapolis: Author.

Ricketts, Cliff. (2003). *Leadership: Personal Development and Career Success (2nd ed.)*. Clifton Park, NY: Delmar Learning.

Texas Farm Bureau. (1993). *Mastering the Media*. Waco: Author.

CHAPTER 7

PAGE LAYOUT: PUTTING IT ALL TOGETHER

OBJECTIVES

After completing this chapter, you should be able to:

- cite the basic layout considerations.
- recognize effective layouts.
- lay out pages using accepted design techniques.
- utilize software as an electronic darkroom.

KEY TERMS

alignment	contrast	proximity
balance	crop	

OVERVIEW

As far as disciplines with agricultural communications are concerned, it could be argued that layout and design is the most important element. After all, it is the last step in building a publication before the public sees it.

Layout and design may not be exclusively the most important, but it shows top billing with all of the other disciplines within agricultural communications, such as speaking, writing, and photography. Agricultural communications students should not ignore the discipline.

The layout and design of a publication is like the front of a home that is for sale. In home sales, "curb appeal" means that when people drive by a house for sale, it looks good from the road and will attract prospective buyers. Just as with homes, newsletters, magazine ads, and other publications need curb appeal to attract readers.

This chapter sets out the basics of applying design techniques to make dynamic and persuasive presentations that generate reader interest and make your publication stand out. In addition, it explores the electronic darkroom and other electronic tools for helping you save time and increase your productivity.

BASIC LAYOUT CONSIDERATIONS

Like any other discipline, layout and design has rules. As mentioned in Chapter 5 on photography, the rule of thirds governs a subject's placement in the frame of the image, and Chapter 3 on writing shows that to create a good news piece, you should use the inverted pyramid style to arrange the information.

The four main elements of design are **alignment**, **proximity**, **balance**, and **contrast**. Each of these elements works in harmony to produce Web designs and page layouts that are dynamic and aesthetically pleasing to the eye.

The four main elements of design help to produce Web designs and page layouts that are dynamic and aesthetically pleasing to the eye.

A way to get a look at these four principles is to pick up any popular magazine or slick color ad. These elements are in use repeatedly in the print media. Look too at the Web sites of big corporations or watch a television commercial. Big businesses have the financial means to hire the best visual designers in the industry, and competition makes them produce ads and content that are on the visual cutting edge.

For the most part, alignment, contrast, balance, and proximity are the principles that govern effective visual presentations no matter what the price. The only difference between the layouts and designs that you see is the content within them. *Content* refers to elements such as the words used, logos, and colors presented. Content is exclusive of the person making the layout, while the design elements are universal.

Alignment

Alignment is exactly what the word means: all of the elements on a page are aligned with one another. The elements can be words, pictures, designs, or logos.

On a basic level, there are six types of alignments. The most basic of alignments are left, right, and center. If you have used any word processing program, you have aligned text or pictures according to these three basic object positions. Taking the three basic alignments, you can make them either vertical or horizontal.

Here are some examples:

Left Alignment

Kicking some dirt from well-sculptured, volcanic-looking prairie
dog mounds, a student of mine asks how deep the
burrows go. As she looks into burrow, other students gather
around to gaze curiously into the six-inch-wide chasm.

Right Alignment

Kicking some dirt from well-sculptured, volcanic-looking prairie
dog mounds, a student of mine asks how deep the
burrows go. As she looks into burrow, other students gather
around to gaze curiously into the six-inch-wide chasm.

Center Alignment

Kicking some dirt from well-sculptured, volcanic-looking prairie
dog mounds, a student of mine asks how deep the
burrows go. As she looks into burrow, other students gather
around to gaze curiously into the six-inch wide chasm.

Which alignment works best depends on the way that the material is used. Play with the alignments, and come up with one that works best for your particular publication (Figure 7-1). Just remember: never mix alignments.

Here is an example of a mixed alignment.

Mixed Alignment

Kicking some dirt from well-sculptured, volcanic-looking
prairie dog mound, a student of
mine asks how deep the burrows go. As she looks
into the burrow, other students gather around to gaze curiously
into the
six-inch wide chasm.

New and Improved!

Come see the new line of 2003 tractors at the area's leading farm and tillage equipment dealership.

Open Monday–Saturday
8:00am–5:00pm

555.555.5555

Or visit us on the web at ***www.ineedatractor.com***

FIGURE 7-1 Sample advertisement using a center alignment scheme.

You can clearly see that a mixed alignment confuses the eye and is cluttered. Let the document you are working on be the guide as far as which alignment would be most successful. A design by committee is often successful because several people are giving their opinions on which alignment is most aesthetically pleasing.

Proximity

Proximity refers to one design object's relation to one another. Objects that are close to one another share proximity; those separated by page space do not.

A good way to think about proximity is to imagine your house in relation to your neighbor's house. By proximity, you live in the same neighborhood, and the houses immediately around you are identified as a singular group when people refer to the part of town you live in.

The same applies to design objects on a page. When objects are placed close to one another, they develop a relationship that makes them appear as a group on a page. A photo and a caption are in proximity to one another because they appear as a group.

FIGURE 7-2   When laying out a page, be sure to keep the elements in close relation to one another. (*Courtesy Russell A. Graves*)

A rule of thumb is to place all objects on a page as close to one another without crowding everything (Figure 7-2). In other words, paragraphs should be grouped close together and not spread out all over a page. On a related note, artwork such as photos should be placed where they relate to the text.

When laying out a page, be sure to keep all of the elements in close relation to one another. Excess space causes the eye to jump around too much, and a document's continuity between elements suffers, making it difficult to read.

Contrast

Contrast is the interaction between one color and another. As you read the words on this page, you will notice that the black letters stand out well against the white background (Figure 7-3).

From an elemental standpoint, contrast is probably the first thing you notice when you look at a Web page or magazine. *High contrast* means that the colors of design elements are nearly the opposite of one another, like

FIGURE 7-3   An example of how contrast affects the way we see pictures and objects.

the color black on white. *Low contrast* means colors are nearly the same shade. Low-contrast objects are hard to read and do not work well when used on printed or Web-based content.

Here are a couple of examples.

This sentence is written in a high contrast color.

This sentence is written in a low contrast color.

You can use contrast to your advantage. As long as you use high-contrast colors, you can vary the shade of different elements such as blocks of text or logos. The point is that not everything has to be the same color. Mix and match colors to draw attention to different elements in the same proximity to one another.

A word of caution, though: Do not overdo contrasting colors because the busyness of all of those shades might make a publication visually confusing. Pick just a few colors, and see how they work with one another.

Balance

Balance is defined as the relationship of one or more design elements and their relation to one another on a page. Not to be confused with proximity, balance seeks to place one design object in relation to another design object so that they complement one another on a page.

Imagine a set of beam scales. If objects weigh the same, the scale is balanced. Add weight to one side, and the scales tip, with one element up and the other down. This analogy is a good way to describe the objects on a page.

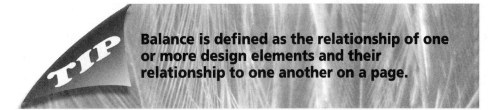

Balance is defined as the relationship of one or more design elements and their relationship to one another on a page.

If the layout is unbalanced, the look is awkward and unprofessional (Figure 7-4). If the layout is balanced, then objects on the page make the presentation aesthetically pleasing (Figure 7-5).

When laying out a page, try to place elements of similar size on opposite sides of the page. For example, one element can go at the top right of the page and the other at the bottom.

FIGURE 7-4   Poor balance.

FIGURE 7-5   Better balance.

If there are two or more design elements, such as a logo and two pictures, then perhaps the two pictures can go at the bottom of the page on opposite sides, while the logo goes toward the top of the page in the middle.

With computer technology, the ability to tweak these four design elements is simple. Be creative. Try different contrast schemes, proximital relationships, design balances, and alignments to see what works best. Although computers do offer some latitude to be creative, the rules of design and layout still apply and should not be abandoned.

USING THE COMPUTER TO SAVE TIME

Computers, their peripherals, and software have made it possible for anyone to produce professional documents for a relatively small amount of

FIGURE 7-6   Computers are a valuable tool for producing professional documents. (*Courtesy USDA/ARS*)

money. The speed and ease of use of computer systems make laying out pages, manipulating photos, and moving and changing text easier than ever before (Figure 7-6).

In the early days of producing documents, each letter had to be meticulously placed on a printing plate and pressed onto paper. Even as electronic publishing began to come of age, blocks of text and various other elements still had to be cut and pasted onto a master document before it was finally printed. Although cutting and pasting a document was still slow, it was a more efficient use of time, even though many masters had to be produced before a suitable layout was reached.

Today's software and high-quality printers have closed the gap between professional and consumer publications. With desktop publishing software, a high-quality, professional layout can be accomplished and printed in a manner of minutes (Figure 7-7).

THE ELECTRONIC DARKROOM

Not too long ago, once a photo was taken, the image for the most part was permanent. You could **crop** a print or make it lighter or darker in the darkroom, but beyond that, the possible changes were minimal.

FIGURE 7-7   Modern printers have the capability of producing high-quality professional ink-jet prints of digital files such as photos for presentation. (*Courtesy Russell A. Graves*)

As computer technology has improved, so has the ability for graphic designers and photographers to manipulate photographic images: They can remove distracting backgrounds, improve contrast, and add elements to make an image that is pleasing to the eye and meets the visual demands of a particular layout.

With computers, traditional images can be scanned while digital images are already stored in the computer. Once in the computer, digitally manipulated images can be placed in layouts and prepared for printing or the Web without having to send original images out for scanning.

Using the Computer to Manipulate Photos

Several software programs allow users to manipulate photos in a variety of ways. The most prominent of programs, Adobe PhotoShop, allows users to incorporate a number of treatments to an image, including removing the subject of an image and placing it on a completely new background. PhotoShop is a powerful tool, and entire books have been written on how to use it.

The discussion here focuses on basic features found in almost all photo editing software. By understanding and mastering some basic photo manipulation principles, you can increase the usability of many of your images.

Cropping

The crop command is used to remove any unwanted portions of an image (Figure 7-8 and 7-9). With the crop command, you can make a horizontal photo a vertical and vice versa and improve the composition of the photo. This command is one of the most useful of all of the manipulations because it allows you to include exactly what you want in a photo.

When using the crop command, keep the rule of thirds in mind and crop accordingly. In addition, crop judiciously; an overly-cropped image looks crowded.

Resizing

When an image is scanned, many scanners set the default images size at a higher resolution than what is needed for Web layouts. A scanned image may be several megabytes in file size.

Since very large files can be an inefficient use of memory resources, it is important that files be resized to conserve Web server space. More important, small-sized photos update from your computer to the server faster and load more quickly when viewed, an important factor to remember because slow-loading Web pages may cause browsers to go somewhere else.

When resizing images, remember that you make them *smaller*, not larger. Therefore, it is always best to scan an image larger than you need and resize it until you get the desired file size. Also, resize in increments so that you do not make the image smaller than desired.

Image Balance

Image balance relates to the colors in an image and their overall relationship to the other colors. Image balance is a conglomeration of factors including the image's brightness, contrast, and gamma. Gamma is the

FIGURE 7-8   Uncropped photo. (*Courtesy Russell A. Graves*)

FIGURE 7-9   Cropped photo. (*Courtesy Russell A. Graves*)

measurement of brightness or darkness of an image on a computer screen. Gamma adjustments are made to images so that the screen image matches the original camera image. In most image editing programs, the user can manipulate these factors to change the appearance of an image. Many of these image controls are much the same as you might see on a television set.

In the image balance controls, you can set the balance of all of the colors or pick one of the primary colors (red, blue, and green) to adjust the appearance of an image.

Brightness

Brightness refers to an image's overall brightness. Making an image brighter or darker also reduces the image's overall contrast.

Contrast

Contrast, as described earlier in the chapter, is the difference between colors in an image. Unlike text, which is usually one color, the contrast of an image is governed by many colors. Therefore, image contrast should be carefully adjusted for optimal image quality.

Rotation

Rotating an image is a common practice to correct an image's orientation as it comes from a scanner. Most image editing programs do the basic rotations such as flip, invert, turn left or right, or mirror. For dynamic designs, try rotating the image a few degrees at a time, and see if you like the results.

AN EXPLANATION OF COMMON FILE EXTENSIONS

At the end of every file on your PC is a file extension. File extensions are usually three-character arrangements of letters that point to the type of file and what kind of program runs the file. Common Web file extensions are .htm, .html, .ram, and .wmf, and common computer files are .exe, .doc, and .xls.

With images, several file extensions exist. Like any other type of computer file types, popularity eventually weeds out competition, and the most popular ones rise to the top. The four most common image file extensions are .jpg, .tiff, .gif, and .bmp. They have unique functions for digital imagery, and each should be used depending on the type of job the picture is to fulfill.

.jpg or .jpeg

Called a "jay-peg," the file extension stands for **J**oint **P**hotographic **E**xperts **G**roup. This format is the most common one used for capturing the minute color and image subtleties for use on the Internet.

.Jpg images are also known as cross-platform, in that they work on any computer regardless of manufacturer and open with a variety of software packages, not just imaging software.

The advantage to .jpg images is that high color definition can be saved at relatively small file sizes. Therefore, images saved in this format take up very little hard drive space and load quickly when accessed from the Internet.

The disadvantages of saving images under a .jpg format are that you cannot make .jpg's transparent for various design schemes or capture extremely high image definition and color depth.

.tiff

Pronounced 'Tiff,' this file extension is best for saving billions of colors in a scanned image. .Tiff images work best for reprinting a scanned image for publication.

Images saved with the .tiff extension keep the full integrity of a scanned image. Those file sizes can be huge—up to 100 megabytes and larger (Figure 7-10).

.bmp

"Bitmaps" are made when an image is converted to bits on a screen. They are useful for erasing parts of an image because the format breaks the overall picture up into small blocks of information.

The advantages of bitmaps are that the image can be easily manipulated with the simplest of image editing tools. The downside of this extension is the large file size needed to map the bitmap.

.gif

.Gif stands for "graphic interchange format" and is pronounced "Giff." .Gif files were among the earliest used Internet image file formats and are still

FIGURE 7-10   Because of the large file size, .tiff files are often saved on CDs. (*Courtesy Russell A. Graves*)

common. Saving images in the .gif format allows you to bleed text or other objects through the image and to animate images at small file sizes.

Gif files have very limited color depth. In fact, a .gif file can represent only 256 colors; the same image saved in a .jpg extension can show millions of colors (Figure 7-11).

The best way to choose a file type is to figure out how the image will be used. Photos for use on the Web or newsletters can probably be handled by

FIGURE 7-11 Clip art, such as this farm scene, is often saved using .gif or .bmp formats.

the .jpg or .gif formats, while images scanned for high-quality publications such as books or magazines need to be saved as .tiff.

Analyze your image needs and educate yourself so you can make the best decision to present your publication professionally.

SAVING IMAGES: HIGH OR LOW RESOLUTION?

Once you have decided what image format best fits your needs, you need to decide if the image should be saved in high-resolution or low-resolution quality. High-resolution images are typically saved at 300 dots, or pixels, per inch (dpi) resolution, while low-resolution images are saved at 75 dpi (Figure 7-12).

Dpi works like this. Imagine that a baseball holds 10 bits of information and a basketball holds 25 bits. Suppose you could put 300 baseballs in a straight line, 20 feet long. The result is 3,000 bits of information in the space. In the same space, you may be able to put only 75 basketballs. Although the basketballs hold a little more volume, the result within the 20-foot span is only 1,875 bits of information. With this analogy, it is easy to see that a 300 dpi image contains much more information than a 75 dpi image does. The result is that the higher-resolution image can be enlarged to larger sizes with greater color depth before any loss of image quality.

The resolution for saving images depends on the application. Small images in newsletters and on the Web can be saved as low-resolution images

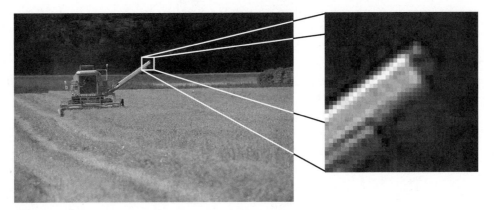

FIGURE 7-12   Tiny dots, or pixels, make up a digital photograph.

because the difference in the resolution is not noticeable. For high-quality digital prints or for an image that will be published in a book, a 300 dpi resolution is the rule.

A word of caution: 300 dpi images work on Web sites. However, because people steal images from a Web site without permission, a high-resolution image can be used easily in an unauthorized fashion. Low dpi images can be taken, but their use outside of the original Web site is limited.

CONCLUSION

With today's computer technology, making impressive layouts is easier than ever before. With amazing speed, you can cut, paste, study various layouts, and play with fresh designs. But although the technology has changed, the principles of layout and design have not. The concepts of alignment, contrast, proximity, and balance are timeless and apply to every successful magazine, television program or commercial, or Web site that you see. Use the rules of design to ensure that time-tested rules apply to make your publication the best it can be.

Computer technology also makes it easy to manipulate images and save them in formats suitable for a variety of needs. Technology has made it possible for even novices to produce professional-looking results.

Although technology continues to expand, the age-old rules of composition, color contrast, and the other design principles never change. Once you master these design concepts, you will see the difference in your publications and so will every one else who views them.

REVIEW QUESTIONS

Multiple Choice

1. _____ is the last step in building a publication before the public sees it.

   a. lay-in
   b. layout
   c. cut and paste
   d. editing

2. Content is _____ of the person making the layout, while the design elements are _____.

   a. universal, exclusive
   b. inclusive, irrelevant
   c. exclusive, relevant
   d. exclusive, universal

3. Left alignment means that _____.

   a. all lines are to the left.
   b. all lines but the first are to the left.
   c. one line is to the left.
   d. none of the above

4. Of all four alignment examples, _____ alignment confuses the eye and is altogether cluttered.

   a. left
   b. right
   c. center
   d. mixed

5. What should be placed where they relate to the text?

   a. captions
   b. photos
   c. news clippings
   d. none of the above

6. From an elemental standpoint, what is probably the first thing you notice when you look at a Web page or magazine?

   a. color
   b. contrast
   c. brightness
   d. layout

7. _____ means to place design objects so that they complement one another on a page.

    a. Imbalance
    b. Coordination
    c. Cropping
    d. Balance

8. What has made it possible for anyone to produce professional documents?

    a. a computer mouse and monitor
    b. computer peripherals and software
    c. a computer keyboard and CD-ROMs
    d. none of the above

9. What command is used to remove any unwanted portions of an image?

    a. cut
    b. paste
    c. crop
    d. cut and paste

10. What does image balance relate to in an image?

    a. colors
    b. contrast
    c. brightness
    d. content

Fill in the Blank

aesthetically	ensure	tools
alignment	publishing	unwanted
balance	colors	Web page
balanced	laying	scanned
brighter	words	300
confusing	cross	logos
crop	thing	low
design	resolution	editing
desktop	brightness	changing
difference	left	file
ease	contrast	busyness
elements	right	crowding

extensions	objects	designs
first	manipulated	imaging
high	manipulating	traditional
invert	mirror	publication
larger	overdo	smaller
layout	page	flip
balance	pictures	rules
design	platform	digital
contrast	professional	reprint
speed	bitmaps	animate
proximity	resizing	bleed
contrast	rotations	composition
presentation	75	magazine
darker	scanned	design
relationship	software	publication
remove	through	brightness

1. _____ and _____ is the last step in building a publication before the public sees it.

2. The four main elements of design are _____, _____, _____, and _____.

3. Alignment is when all of the _____ on a page are aligned with one another.

4. Elements can be _____, _____, _____, or _____.

5. Place all _____ on a _____ as close to one another without _____ everything.

6. _____ is probably the _____ _____ you notice when you look at a _____ or _____.

7. Do not _____ contrasting colors, as the _____ of all of those shades might make a publication visually _____.

8. _____ is defined as the _____ of one or more _____ elements and their relation to one another on a page.

9. If the layout is _____, then objects on the page will make the _____ _____ pleasing.

10. The _____ and _____ of use among modern computer systems make _____ out pages, _____ photos, and moving and _____ test easier than ever before.

11. With today's _____ _____ software, a high-quality, _____ layout can be accomplished and printed in a manner of minutes.

12. With computers, _____ images can be _____ while _____ images already lie in the computerized realm.

13. The _____ command is used to _____ any _____ portions of the image.

14. You can improve the _____ of a photo by cropping.

15. When _____ images, remember to make them _____, not _____.

16. _____ refers to an image's overall _____.

17. Making an image _____ or _____ also reduces the image's overall contrast.

18. _____ is the _____ between _____ in an image.

19. Most image editing programs will do the basic _____ such as _____, _____, turn _____ or _____, or _____.

20. The four most common image _____ _____ are .jpg, .tiff, .gif, and .bmp.

21. .Jpeg images are also known as _____-_____ in that they work on any computer regardless of manufacturer and they open with a variety of software packages, not just _____ _____.

22. .Tiff images work best when you need to _____ a _____ image for _____.

23. The advantages of _____ are that the image can be easily _____ with the simplest of image _____ _____.

24. .Gif format allows you to _____ text or other objects _____ the image and to _____ images at small file sizes.

25. _____-_____ images typically saved at _____ dpi resolution while _____-_____ images are saved at _____dpi.

26. When using the _____ of _____, you can _____ that time-tested rules apply to make your _____ the best it can be.

ENHANCEMENT ACTIVITIES

1. After completing a feature article from Chapter 3, illustrate it with photos using the lessons learned in Chapter 5. Then, using a computer, scan the images, and lay them out in a fashion that follows a magazine layout and follows the rules of balance, proximity, and alignment.

2. In preparation for Activity 1, crop and adjust all artwork in the computer so that it fits the design scheme for the magazine layout.

3. After scanning an image, save it under different file formats as prescribed in the text, and note the differences.

4. With the aid of your classmates, assemble all layouts into a single, cohesive publication. Once the publication is made, present it to teachers in the school, school administration, and agribusiness leaders throughout your community.

WEB LINKS

To find out more about the topics discussed in this chapter, visit these Web sites:

Successful Web Pages: What Are They and Do They Exist?
<http://www.intertwining.org/>
   (Search terms: collaboratory papers)

Visual layout and elements
<http://www-3.ibm.com/>

Typography and page layout
<http://www.dtp-aus.com/>

Image formats
<http://www.howstuffworks.com/>

Alternatively, do your own search at <http://www.alltheweb.com>
   (Search terms: photo editing, photo manipulation, layout and design, image manipulation, file extensions, digital photos)

BIBLIOGRAPHY

Hartenstein, Shannon. (2002). *Preparing for a Future in the Agriculture Communications Industry*. Manhattan: Kansas State University.

Texas Tech University. (2003). *Ag Communications CDE Preparation Guidebook*. Lubbock: Author.

Williams, Robin, and Tollett, John (2000). *The Non-Designer's Web Book* (2nd ed.). Berkeley: Peachpit.

CHAPTER 8

WORKING AS A FREELANCER

OBJECTIVES

After completing the chapter, you should be able to:

- define the term *freelance journalist.*
- outline a market analysis for freelance work.
- identify possible freelance markets.
- know how to craft a query letter.
- recognize potential pitfalls as a freelancer.
- write to magazines requesting freelance guidelines.
- articulate rights issues as they apply to freelance work.
- cultivate and maintain professional freelance relationships.

KEY TERMS

exclusive rights	freelance	query letter
first-time usage	masthead	unlimited usage

OVERVIEW

The freelance market for writers and photographers is a growing one. Many companies that produce media content are seeking talented people who can provide services on a contractual basis. The reason is simple: they pay when they need work done and do not have to keep an extra employee on their permanent payroll. For many company's business models, it is cheaper for them to pay a freelancer $500 on a periodic basis than to pay a regular employee the same amount every week. In addition, they can hire some of the most talented people in their respective fields to work on an as-needed basis.

As long as there are media outlets, there will always be a need for free-lancers. Magazines, for example, do not have the budget to have people all over the country on their regular payrolls. It does not make financial sense to have bureaus in every state and have writers producing stories every month when only a half-dozen will be put into print. Instead, it is smarter for them to hire people only when they need them. If a story breaks in Missouri, they can hire a freelancer in the area to cover the story (Figure 8-1).

This chapter explores the rewards and challenges of being a freelance writer or photographer. It explains how to work with editors and analyze the competition, and what to look for in a contract. Finally, it sets out how to make contacts with editors that highlight your skills and professionalism in a way that does not tread on their busy schedules.

In the end, you will have the basic knowledge that it takes to start down the road to becoming a freelancer—basic knowledge that lets people know that you are willing to do professional work. In return, you expect to be treated like a professional in both conduct and pay.

FIGURE 8-1   In this chapter, you will learn how to approach editors about stories on such subjects as biotechnology. (*Courtesy Russell A. Graves*)

WHAT IS A FREELANCER?

You have the talent and the ability to produce quality writing or photography, or both. You have written for your school's newspaper and your local newspaper, and even provided some content for Web sites. Your work is appreciated by the people you have met and is in demand.

If you are asked regularly to provide visual or literary content to various media outlets, you probably have what it takes to sell some **freelance** work. Many times, when people get started doing work, they are happy to see their name in print in exchange for payment of services. That is fine. However, if the demand for your work gets to the point where you spend a considerable amount of time to produce content for other people, you should declare yourself a freelancer and begin charging for your services. As the old saying goes, "Time is money," and you should consider your time as valuable as anyone else's.

TIP

If your work is in demand, it may be time to consider yourself a freelancer and charge for your services.

What exactly is a freelancer? A freelance journalist is one whose declared allegiance is not with one company. Freelancers work for hire and may have several clients to whom they provide material on a regular or a semiregular basis. Their talents are not dedicated to a single company, so they are free to shop them around to whomever they choose (Figure 8-2). They are free to negotiate pay scales and contracts based on their abilities and can walk away from work if they decide that it is not right for them.

It can be rewarding to work for yourself, but there are some challenges that you may face, such as competition, producing work on a timely basis, and coming up with fresh ideas. The challenge is not impossible. Many people make a living as freelance journalists, but it takes dedication, sacrifice, and discipline.

THE FREELANCE MARKET

Just about every kind of industry has a freelance segment that accompanies it. There are freelance carpenters, welders, stonemasons, business consultants, and others although "freelance" seldom precedes their working titles.

FIGURE 8-2   As a freelancer, you are free to shop your talents around to anyone you choose. (*Courtesy USDA/ARS*)

The media industry is no different. Within the media industry, there are freelance editors, illustrators, advertising representatives, art directors, writers, and photographers. Magazines often hire out services on a regular basis because retaining staff positions in these fields may not be financially feasible for them. Therefore, to make sure they have an abundance of top-quality work, magazines, newspapers, and Web sites hire freelancers—and hire them often. In fact, some small magazines have a complete freelance staff.

TIP   Due to limited budgets, magazines often hire the services of freelancers.

Since the focus of his textbook revolves around writing and photography, we focus on those two professions.

WHAT KIND OF FREELANCE MARKETS EXIST?

For freelancers, at least four distinct types of markets exist: magazine, newspaper, Internet, and advertising markets. There are more markets, but these are perhaps the most common. Within each market, submarkets can be found. Each market has its own set of rules as to how it works with free-lancers, and there are benefits and disadvantages to working with each one. For writers and photographers however, plenty of places exist in which freelancers can sell their material (Figure 8-3).

If you are purely a writer, markets for your work abound. You need look no further than the local newspaper for many opportunities to sell work (Figure 8-4). Although the pay usually is not high, newspapers can help you cultivate your style and tighten your writing. Web sites are look-ing for content just as newspapers are. With the explosion of the Internet as a source of information, good writers are necessary for Web entrepre-neurs. Like newspapers, the pay is not very high, but this work will give you a chance to perfect your craft and sell plenty of content if you make enough contacts within the industry.

Photographers face the same challenges and might want to consider taking the same route as writers in getting started. Local newspapers are probably the easiest to break into, followed by Web sites; local, regional, state, and then national magazines next; and advertising markets, perhaps the hardest markets to break into.

SUNDAY, MARCH 2, 2003 - 12 PAGES - 50¢

nted:
il will decide this
police chief issue

even paid and 28 volunteers. re are no openings for volun- s.

he council voted unanimously to all water and sewer lines to ldress Regional Medical ter's new 16-physician clinic,

retire.

Royce Seibman discussed the RC and D Rural Leadership Program and requested help from the council in identifying persons who could participate in leadership training. The council will contribute $250

FIGURE 8-3 Newspapers are a potential market for freelancers. (*Courtesy Childress Index, Inc.*)

Farm Service Agency County Committee C

The Farm Service Agency (FSA) is seeking candidates for the locally led County Committee election to be held this fall.

"It is crucial that every eligible agricultural producer take part in this election process because County Committees serve as a direct link between the agriculture community and the U.S. Department of Agriculture," says Kathy Williams, Childress County Executive Director. "The County Committee system needs everyone to get involved-from voters to com-

FSA Cou local guida on issues price supp establishm and marke conservati indemnity, for some disaster pr Committe eyes, ears ranchers n 60 years.

"The Co

FIGURE 8-4 Local newspapers are perhaps the easiest to break into. (*Courtesy Childress Index, Inc.*)

Pay

"What does being a freelancer pay?" That is one of the most commonly asked questions by many looking to get into the business. The usual answer is, "It depends . . ." It depends on a number of factors including the quality of your work, your personal financial goals, the number of hours you are willing to put into the business, and your marketing expertise.

Being a freelancer is like any other job. Some freelancers are stars in their profession, and others are happy at just getting by. You need to determine where you want to be should you decide to become a freelancer.

Pay depends on the quality of your work. Like a baseball player, it is hard to start in the major leagues of the freelance business and sell work to all the major markets from the start. Instead, most freelancers begin in the minor markets, such as local newspapers. There is no dishonor in starting small. Local markets are an excellent way to perfect your craft in order to step to the next level (Figure 8-5). How will you know when you are ready? As soon as you start submitting work to larger markets and they begin accepting it, you are ready. In short, it is the markets you work for that ultimately decide how good your work is.

The pay for freelancers often depends on the quality of their work.

How much you make also depends on your financial goals. Some want to make working as a freelance writer or photographer a full-time job and do well. Others are happy supplementing their household income from another job with work as a freelancer. Ultimately, determining your financial goals is a personal endeavor. When making financial goals, however, be realistic: Few freelancers make hundreds of thousands of dollars annually. Do plenty of research in the field, and talk to those who are successful. They can tell you how much you can expect to make.

How much money you will make as a freelancer depends on your financial goals.

FIGURE 8-5   Which picture do you think is the better one? As you become an established freelancer, the market will help you decide which picture is better based on market needs. (*Courtesy Russell A. Graves*)

On the subject of how much the various markets pay, it typically depends on how many people will see your work once it is in print. For example, newspapers are typically the lowest paying of the four markets. Local, small town newspapers may pay only $10 for an article or a photograph. Big city newspapers may pay a little more but not appreciably so. This is not to say that newspapers are a bad idea to work for. Most newspapers are printed on a daily basis and their advertising rates are much lower than those of magazines. They must stretch the editorial budgets over a much longer time span than a magazine that prints only once a month.

Web markets are next on the rung of pay scale in the four markets. Since the overall revenue that Web sites take in the form of advertising is, on the average, lower when compared to print media, pay for written or photographic content is typically lower than what magazines pay and slightly higher than newspapers. For example, a Web site may pay $50 for a 1,000-word article and $25 for each photo used. Like newspapers, the pay can vary over what is quoted in this book. Like newspapers and magazines, the more popular the Web site is, the more revenue it demands for advertising space and the more it pays for freelance content.

Magazines are the pinnacle for many in the freelance business. Many people make a good living as full-time freelancers for a number of different magazines. The key is to find a core group of magazines that value your work and are willing to pay you fairly for what you do.

Magazines are notorious for having a very wide pay scale. Some small-circulation magazines do not pay anything, while some of the very largest may pay $2,500 for a cover photograph alone. Typically, though, magazine pay rates revolve around their paid circulation numbers. For example, a magazine that has a modest circulation of 250,000 copies a month may pay anywhere from a quarter to fifty cents per word for articles and, depending on the amount of a page they fill, anywhere from $50 to $300 for an inside photograph and $500 for a front cover.

Advertising agencies pay the most for freelance work. Many ad agencies do not work one on one with freelancers. Instead, they work with agents who often represent a number of writers or photographers. Although the opportunities with ad agencies are small in comparison with other types of media outlets and the competition fierce, doing work with ad agencies can be financially rewarding. For example, it is not unusual for a single image from a photographer's file to bring several thousand dollars when it is contracted for exclusive use in an advertising campaign.

Although markets vary, a good strategy is to do a little in each of the markets while specializing in one. In that way, as in production agriculture, you diversify your clients, thus ensuring a constant cash flow into your business.

Always find out what a potential client pays before doing work with it. Both parties involved will benefit when each side has a complete understanding as to what is expected of them.

Your Rights as a Freelancer

As a freelancer, you have certain professional rights. Just because you are working for a publication does not automatically mean that the work you are producing belongs to it. In most cases, the work produced is leased to a magazine—for example, for a specific length of time and under specified usage terms.

The most common type of usage rights that most freelancers grant is **first-time usage** rights. Sometimes referred to as one-time usage rights, these rights are, in effect, a lease agreement. For example, if you write a magazine article for a beef publication and you grant it first-time usage rights, that means that it gets to use the work for the first time and for a specific length of time. It also means that it may use the article once under the terms of the usage rights. Since most magazines are published every month, the length of time usually spans the time between issues. When first-time usage rights are granted, the work still belongs to the freelancer. After the specified time has passed, you are free to resell (or lease) the work to anyone else who will pay you for it. The same usage rights apply to the work whether it is a photograph or an article.

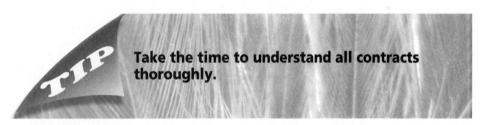

Take the time to understand all contracts thoroughly.

Although articles are seldom returned because most writers keep an electronic copy of their work, photographs are always returned under first-time usage contracts. Many photographers report selling the same image repeatedly through the years and gaining a substantial income from the photo. Typically, first-time rights pay the least amount of all of the usage rights negotiated with a publication.

Unlimited usage varies a bit from first-time rights. With unlimited rights, the user of the content can reproduce the photograph or the article more than once. Like first-time rights, a time limitation is specified. For example, a cotton magazine may want to use an image you took for its table of contents or logo (Figure 8-6). With unlimited usage, it can reproduce the

FIGURE 8-6   A cotton magazine may look for specific uses for its photos and, depending on the usage of the image, the rights granted to the magazine will vary. (*Courtesy Russell A. Graves*)

image as many times as it wishes for the length of time specified in the contract—often a one-year period. No one else may use the work during the specified time frame. Since these rights are more restrictive to the freelancer, a price higher than first-time usage rights should be negotiated.

The most restrictive rights contract that a freelancer enters into is **exclusive rights**: the entity purchasing the rights has the exclusive right to publish and republish the work as it sees fit for an indefinite period. Often, exclusive rights are the same as if the company had bought the work outright because it controls how the work is used and can reuse it at its discretion.

Exclusive rights in the agricultural communication magazine industry are rare and are negotiated only under very unusual circumstances. They are often purchased in the advertising business because corporate identities are often tied to the images or copy presented in magazines, on television, or on billboards. It should be noted that most rights negotiated with photo users generally do not require the services of an attorney. However, for deals that involve several thousand dollars, legal representation may be advisable.

Exclusive rights are restrictive to the freelancer, and only the highest prices should be negotiated. Once the work is cloaked under exclusivity, a freelancer's potential to earn income from it through other sources is limited or nonexistent.

MAKING CONTACT WITH THE MEDIA

You have educated yourself on what being a freelancer is about. You learned the pay scales and understand contractual terms, and now you are ready to wade into the freelance markets. What is next?

Like most enterprises, protocols are in place that guide people on how to do business with places that contract freelance agricultural journalists. From this point forward, we will concentrate on doing business with magazines since they typically use more freelance work than do newspapers (which are largely staff written) and ad agencies (work through agents).

The Masthead

If you look on the first few pages of a magazine, often on the Contents page or the page immediately after the Contents, you find the **masthead** (Figure 8-7). The masthead contains the information you need to contact a magazine about the prospect of freelance work.

The masthead lists everyone who works for the magazine and their mailing addresses, the phone and fax numbers of the magazine, e-mail addresses, and subscription information. Often, the chain of command for the magazine starts at the top and works its way down. For example, typically the top name on the masthead is the publisher. Think of the publisher as the chief executive officer of a company. This person is in charge of the day-to-day business and personnel issues within the magazine and probably deals very little with editorial content. On the masthead, look for the editor's name if you are interested in submitting proposals for articles and the photo editor's or creative director's name if your aim is to license the magazine your photographs.

Once you have identified the correct person to contact, write the person a letter or send an e-mail. If you have no prior working relationship with the editor, do not call. Like other professionals, editors are busy people who may have deadlines to meet and may not have time to chat on the telephone. The most unobtrusive way to contact them is with a letter or e-mail. Send a self-addressed, stamped envelope so they can answer it when they have time to give it their full attention and can send a reply without costing the publication any expense.

Staff

Editor/Publisher
Kent Lynch

Production Coordinator
Art Vasquez

Technical Advisor
Robert Anderson

Dally Times (ISSN - 10861831, USPS - 013616) is published monthly. Published by Dally Times Inc. The publication office is located at 4748 Hwy 377 South, Fort Worth, Texas 76116. Periodicals postage paid at Fort Worth, Texas.

Official publication of:
Original Team Roping Association, American Cowboys Team Roping Association and the Senior Team Ropers Association.

Subscription Price: 12 months-$20.00.
Visa/Mastercard accepted.

DEADLINE
FOR EDITORIAL AND ADVERTISING IS THE 10TH OF THE MONTH PRECEDING PUBLICATION DATE.

Material may not be reproduced without permission from publisher. Dally Times hereby expressly limits its liability resulting from any and all misprints, errors and/or inaccuracies whatsoever in advertisement or editorial content, to the refund of specific advertisement payment and/or the running of a corrected advertisement or editorial correction notice. The opinions and views expressed in all editorial materials are those of the writer or the person interviewed and not necessarily those of Dally Times.

POSTMASTER:
Send address changes to:
Dally Times
4748 Hwy 377 South
Fort Worth, TX 76116

Dally Times

FIGURE 8-7 You can find a magazine's contact information by looking on the masthead. (*Courtesy Daily Times, Inc.*)

Payment

We pay on publishing. Payment is $600 for a full-length of 2,000 words or 30 cents a word. We purchase one-time rights to all photographs. Photo rates are $1000 for a cover while inside color photos rates range from $80 to $250. A Social Security Number must be included by first-time contributors before a check can be issued. All work is paid for upon publication.

FIGURE 8-8 Editorial guidelines give a freelancer an idea of submission policies, pay rates, and other important information.

What to Ask For

In your first letter to an editor, get to the point right away. The first letter is not the time to outline your journalistic philosophy or provide your educational or work background. An editor who wants that information will ask. Ask for a copy of the magazine's editorial guidelines for freelance writers or photographers and the publication's pay scale for freelancers (Figure 8-8).

These guidelines provide specific information on the editorial focus of the publication, the process for submitting proposals, and the correct protocol for submitting photos and articles. For example, for photographs, many magazines want only 35mm and larger slides, not prints, submitted.

They may specify that a maximum of 40 slides be sent at any one time and that the slides be sent in clear plastic pages for viewing. They may also request that all slides be mounted on cardboard or plastic and that all caption information be put on the slide mount.

The guidelines specify how to submit articles. Nearly all magazines want articles double-spaced. Some may request the submission be printed and stapled, while others ask for electronic copies on disk or by e-mail as an attachment.

Pay varies by publication. Some offer no pay at all, while others outline exactly what they pay and whether they pay when a work is accepted or published. With articles, some magazines pay a flat rate for a complete package, while others pay by the word, with the rate varying from five cents to around fifty cents per word.

For photographs, some magazines pay only one rate for one-time usage of photos regardless of the size that a photo appears in the magazine. Other magazines have space-specific rates and pay for the size that the photo may appear in the magazine. The size is usually measured in quarter-pages. For example, less than a quarter-page photo brings the lowest rate, and three-quarters to a full page or larger brings a higher rate. Front covers pay the highest rate per photo usage among most magazines.

These guidelines are your first contact with a magazine for which you may choose to work. By following the magazine's protocol, you establish a professional relationship that tells the editor that you can follow directions and are willing to accept its way of doing business. Once you have decided to take the next step, it is time to construct a query letter.

WRITING THE QUERY LETTER

A **query letter** is your first link to breaking into the freelance business. It works as your representative, and its purpose is to earn you an assignment from the publication you want to work for. A query letter must be well written in order to instill an editor's confidence in your abilities; it must be to the point so that you do not unduly waste a busy professional's time; and it must clearly sell the angle in which you are proposing the article so the editor can see if there is a fresh idea in what may be an old subject—all in a single page. Although it sounds tough, once you know the parts of a query letter and how to format one, it soon becomes second nature.

An Example Query Letter

Following is an example of a query letter that earned an assignment for the author:

February 21, 2001

Editor
Agriculture Magazine
1211 North Main
Anywhere, USA 22222

Dear Mr. Editor:

In the cattle industry, we are constantly fighting a battle to control parasites. Internal parasites rob our cattle of valuable nutrients that hamper feed efficiency and milk production. Consequently, most cattle raisers identify a worm infestation as a problem. However, external parasites often go untreated.

According to recent studies by Texas A&M University, external parasites rob cattlemen of millions of dollars annually. Flies affect cattle by reducing average daily weight gains in calves, milk production in cows, and the spreading of diseases such as pinkeye and anaplasmosis. Lice and grubs damage hides and irritate cattle to the point where all they can think about is scratching instead of eating. Therefore, the potential for profit loss is obvious.

The article that I am proposing will cover some basic management practices that cattlemen can follow in order to control external parasites. In the article, I plan to cover topics such as the insecticide tag that should be used according to the time of year, the use of back rubbers and how to make them more effective, and how to help reduce your chances of having a chemical resistant population of flies in your pasture.

As you can see, this type of article would be very beneficial to the readers of your magazine as it hits home with most, if not all, of your subscribers. I would appreciate your consideration on this matter and if you have any further questions, you may contact me at the phone number or address below.

If you wish to see any copies of my previous work, please feel free to ask for it.

I look forward to hearing from you soon.

Sincerely,

Russell A. Graves
1000 Main Street
Anyplace, Texas
555.555.5555
Russell@russellgraves.com

The Parts of a Query Letter

A query letter should be formatted just as any standard business letter would. All of the text is aligned on the left side of the page with no indentions and is single-spaced—except between salutations, closings, and paragraphs.

At the top of the page is the date the letter was written. It should always be in long form for a formal look (that is, February 21, 2001, rather than 2/21/01).

Next, put the name of the person you are writing to, followed by the name of the magazine and the magazine's address. After that, write the salutation. For the first-time contributor, start the letter with Dear Mr. or Ms. Editor, with a colon at the end of the person's name. Once a professional relationship is established, then you may feel comfortable addressing the editor by first name.

After the salutation comes the body of the letter. The first paragraph states your proposed topic in one or two sentences. Remember to get to the point, and make the sell quickly. The next two paragraphs provide background information on the subject so that a case for running the article is solidly argued. The fourth paragraph spells out the angle in which you will cover the subject so that the editor can determine if your idea is a fresh one. The fifth and sixth paragraphs wrap up the letter on a positive note and leave the door open for the editor to contact you in case he or she has any questions or wants to see samples of previously published work.

Close the letter with the formal word, "Sincerely."

Following the closing, drop four spaces down to provide space to sign your name and then type your name, followed by your address, phone number, and e-mail address.

Once the letter is written, print it on letterhead or heavy bond paper for a professional look. Use your computer's word processor to print the name and address on the outside of the envelope you are sending. It makes no sense to create a professional query on letterhead only to handwrite the editor's name on the outside.

Finally, include a self-addressed stamped envelope in the mailing for the editor's convenience in responding to you.

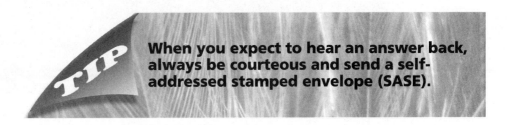

When you expect to hear an answer back, always be courteous and send a self-addressed stamped envelope (SASE).

EDITOR–FREELANCER RELATIONS

Like personal relationships, business relationships, such as the one between you and an editor, are based on mutual trust (Figure 8-9). You know that the editor will respect your work, and the editor knows that you will always conduct yourself in a professional manner. A good working relationship can be fruitful for many years. Although the editor you work with may change jobs, your reputation for quality work stays behind at the magazine. The next person to step in will have heard about your creative abilities and will want to cultivate a relationship with you.

A good relationship with one publication often brings in business from other publications. Editors often share names of good freelancers with other editors with whom they have professional relationships. This word of mouth can also spill over into other freelance market opportunities, such as advertising work.

Building a Strong Relationship

How did you build a strong relationship with your best friend? Although you did not necessarily plan to become best friends, things just fell into

FIGURE 8-9   Like personal relationships, business relationships are based on personal trust. (*Courtesy USDA/ARS*)

place over time. Not surprisingly, the same factors that made you best friends with someone are also the same dynamics involved in cultivating successful business relationships. Personal friendships take years to cultivate but in the fast-paced business world, most people trying to earn a living do not have that long to wait. Here are some tips for helping you to build strong editor-freelancer relationships quickly and solidly.

Do What You Say You Are Going to Do

This point is perhaps the most important thing you can do to build a successful business relationship. If you agree to a deadline of February 1 on a feature article, make sure you hand in the work as early as you possibly can. That way, if an editor is not happy with your coverage of the subject or finds grammatical or factual mistakes in a piece, you have time to make changes or corrections. By missing deadlines or not doing an assignment at all, you put the editor in a bind. He or she has to scramble to find something to fill the space reserved for your work.

Missing deadlines can be deadly to your freelance career, and it is best not to make it a habit. Just like word of your good-quality work travels around in editorial circles, so does the negative news.

Keep a calendar, and get your work done on time.

Be Prompt in Responding to Correspondence, and Communicate Often

If an editor calls, writes, or e-mails you, respond as quickly as you can. Editors work on tight deadlines, so if they contact you looking for information, they need it in a timely matter. Make it a habit of getting back to the people who contact you.

Be prompt in returning phone calls.

Also, make a habit of contacting editors in the middle of an assignment they have given you. That way, you can assure them that everything is on track and that you will meet the deadline.

Do Not Overstate Your Abilities

When you are starting out, it is tempting to overstate your expertise and what you may be capable of. Resist the temptation. In the early negotiations for an assignment, state exactly what you are capable of doing and when you can do it. If you are asked to take on an assignment that you feel you cannot adequately complete, let the editor know, and state the reasons. The editor may be disappointed but surely will respect you for knowing your limitations and making them clear.

More often than not, most assignments are within your ability to complete, and editors who appreciate frankness and honesty will call on you for your expertise.

Know the Terms of the Contract Early in the Game

Once an assignment is imminent, make sure you know what the terms of the contract are. For example, how long does the piece have to be? How much is the pay? Will the check be sent on acceptance of the piece or on publication? What kind of usage rights does the client want to buy?

Understanding the contract protects both parties (Figure 8-10). First, you are aware of all of the terms of the agreement and can respectfully decline the offer if the terms do not meet your expectations. Second, the publication is protected because once you agree to a set of terms, you cannot go back to the table for more pay if the article took you longer to write than you anticipated.

Respect an Editor's Expertise

Editors are professionals in their fields and should be treated as such. One of the hardest yet most productive things you can learn to do is accept constructive criticism (Figure 8-11). If you work for a number of publications, you will discover that many of them have their own distinct style that they like their articles written in. With varying editorial styles, magazines can set themselves apart from competing publications.

An editor's suggestions for making a piece of work read better or fit the magazine's editorial style better is not a personal jab; it is a suggestion for

1. The Author hereby grants the Publisher, its successors and assigns, during the term of copyright including renewals thereof the exclusive license to print, reprint, reproduce, publish and sell the work in book form in any language any place in the world.

2. The Publisher shall, within one year after the Author has delivered a complete and satisfactory manuscript as provided in paragraph 4, publish the work at its own expense and in such style and manner and at such price as it deems suitable. In no event shall the Publisher be required to publish a manuscript which in its opinion violates the right of privacy of any person or contains libelous, obscene or other unlawful matter. The Publisher shall not be responsible for delays caused by any circumstance beyond its control.

FIGURE 8-10   Take the time to understand all contracts thoroughly.

making you a better journalist, and you should view it in that light. Egos should be left behind and an open mind brought to the table when your work undergoes scrutiny by an editor.

The road goes both ways too. If an editor disagrees with content you know to be factually correct, respectfully disagree and state your case. That way, you both work together for concrete resolution of an editorial issue, and you have done it as a team—the way fruitful relationships of all kinds should resolve conflict and, ultimately, grow.

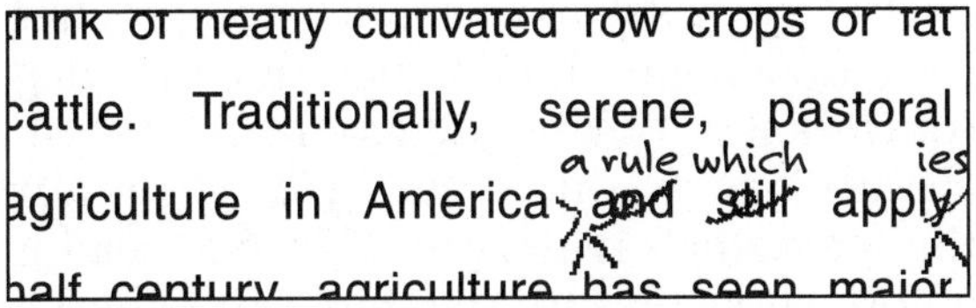

think of neatly cultivated row crops or fat cattle. Traditionally, serene, pastoral agriculture in America, and still apply *a rule which ies* half century agriculture has seen major

FIGURE 8-11   As a freelancer, you should learn to accept constructive criticism.

MAKING A LIVING AS A FREELANCER: SOME CONSIDERATIONS

The Pay

The pay of a freelancer varies greatly and is dependent highly on the individual's skills and fortitude. Successful individuals often work many hours and sacrifice personal time in order to be one of the top freelancers in a given market. These people have developed reputations for getting the job done in a timely manner and producing high-quality work.

TIP   **The pay for a freelancer often varies.**

Success, however, is a subjective term. What success is for one person does not always apply to others. Consider pay, for example. How much pay would it take for you to consider yourself a success? How about the person sitting next to you? Do you think the two of you would give the same answer? Probably not.

Making a good living as a freelancer means you have to be prolific in producing material. Suppose you get an assignment for an article and photo package that pays $500 plus expenses. It sounds good on the surface, but in order to make a modest $30,000 annually, you have to earn sixty such assignments—that averages five per month.

That is not saying it cannot be done, but it is a tough business in which to earn a windfall of money. Many freelancers supplement their editorial income by other means, such as having a full-time job in one field and doing freelance work as a sideline enterprise.

In order to determine the pay for a magazine, send a letter or e-mail requesting the latest copy of its writer's and photographer's guidelines. In the guidelines, the document should spell out the magazine's editorial focus, the type of work it is looking for, and its pay rates.

Finding the editor's name and e-mail address for a magazine is easy. First, look for the magazine's masthead. The masthead is where you will find information on who works for the publication and in what capacity. Usually the masthead is found on the page after the Contents page.

Study the masthead carefully, and make sure you find the editorial address for the publication. Many times the masthead also lists the advertising and subscription offices as well. When you send a letter to the editor at the appropriate address, always send a self-addressed stamped envelope (SASE) so that you pay the cost of the reply. By including an SASE, you increase your chances of getting a reply.

Making Ends Meet

Like any other business, the freelance writer's and photographer's business has its shares of pitfalls and rewards. One pitfall is cash flow and, ultimately, making ends meet. When you talk to freelancers, you will find that one of their toughest challenges does not come in the field; it is in the office and at their bank (Figure 8-12).

Staying in the freelance business requires careful planning and management of finances. The reason is that work done today may not be paid for several months. Many magazines list in their guidelines that they pay on publication of the work. That would be fine if your work were published just a couple of days after you had completed it. That is not the case. Most publications work at least three months in advance, and some work even a year in advance. Therefore, work done in May might not be paid until December.

Ultimately, you end up waiting for money in order to keep your business running. The most successful freelancers recognize this limitation and plan accordingly so that they have a constant source of business funds available to pay bills, salaries, and associated business and production

6	(3) 46, (3) 50	$	480.00
Total Due For All Photos Used		**$**	**480.00**
his signed invoice			
	2/15/01		

FIGURE 8-12   Staying in the freelance business requires careful financial planning and management.

expenses. Their plan may include booking enough work so that checks for work already done are coming in the mail on a regular basis or using a line of short-term credit from a bank or other lending establishment.

The way to make ends meet is to develop multiple streams of income. As a freelancer, you have the unique ability to work for whomever you want. Therefore, it is important to find business wherever you can get it. For example, look to different magazines for opportunities to work. In addition, book publishers, local newspapers, Web sites, advertising agencies, radio stations, and hosts of other entities are often looking for people who can work on an as-needed basis. The key is not to lock yourself into any one revenue-generating source. Get creative, and explore all of your options as they match your abilities.

TIP To make ends meet, many freelancers must rely on multiple streams of income.

The Competition

Freelance journalists face competition that is sometimes very tough. With the advances in computer and photographic technology, the gap between professional and consumer electronic devices has closed considerably. As a result, virtually anyone is capable of producing professional Web sites, editorial copy, and artwork with a greatly reduced learning curve as compared to only a generation ago (Figure 8-13).

The technological gap between all people has narrowed to the point that anyone with the inclination to become, say, a photographer, can do so without any formal training. People all over the country are now producing work that is of high enough quality to be in the freelance market, but they may choose not to pursue that avenue. Those who see value in their work and seek to exploit its monetary value represent your competition.

Compared to a readily pursued photography market such as wildlife photography, the agricultural market is much smaller and more specialized (Figure 8-14). Most likely, there are fewer people pursuing money-making opportunities within the agricultural publications arena than other common freelancing ventures. Although the numbers may be smaller, the competition is just as keen.

FIGURE 8-13   With the technological gap closing, it is possible for anyone with some skill to take publishable photographs. (*Courtesy Russell A. Graves*)

Competition is not necessarily a bad thing. It can drive you to become better at anything you do if you are willing to face the challenges. The way to beat the competition, especially in the agricultural journalism game, is to produce quality work consistently and develop fresh ideas with new angles.

Staying Fresh

With competition being a constant part of doing business, the need to stay fresh is ever-present. Staying fresh means developing new ideas while looking out for fresh approaches to old ideas. That rule applies to both photographers and writers.

To stay fresh, you must always stay on top of the specialty you choose to pursue. For example, if you plan to cover the cotton industry exclusively, you need to know about international trade issues, farm bill legislation, and emerging technology within the industry. The only way to stay current is to read industry news releases and insider reports from the industry.

If you are a photographer, study advertisements and magazines. Trends in photography come and go, and you need to stay on top of the trends.

FIGURE 8-14   Although both outdoor subjects, the wildlife and agricultural markets have different levels of competition. (*Courtesy Russell A. Graves*)

Granted, agricultural media typically use straightforward shots of crops or livestock. Nevertheless, the ability to produce a different look to the same old subject will undoubtedly keep you and your work marketable.

CONCLUSION

Freelancing is not for everyone. It is a challenging way to be an agricultural journalist and is characterized by many peaks and valleys. The main challenge is to move from being an unknown writer or photographer to one who is in demand from markets all over the country.

The key word though is *challenges*, not *impossibilities*. With the right knowledge, an aspiring agricultural communicator can become a freelancer on a full- or part-time basis with success. That knowledge includes knowing the protocols for making contact with magazines and how to write successful query letters so that you can land assignments.

Being a successful freelancer also means knowing your rights when it comes to the work you produce and how much publications pay for various uses of creative work. Being a successful freelancer means that you know how to cultivate and maintain successful working relationships with the people to whom you provide creative content.

REVIEW QUESTIONS

Multiple Choice

1. A freelance journalist

   a. works for a specific publication.
   b. contracts work to only one publication.
   c. works for hire.
   d. gives his or her work away for free

2. According to the text, how many common freelance markets exist?

   a. 1
   b. 2
   c. 3
   d. 4

3. Newspapers can

   a. never help your freelance career.
   b. help cultivate your style and tighten your writing.
   c. help give advice and structure to your career as a freelance writer.
   d. help circulate your name to prospective publications.

4. Magazine pay rates revolve around

   a. how good your writing or photographs are.
   b. the amount of times you have done business with them.
   c. your reputation as a freelancer.
   d. their paid circulation numbers.

5. Advertising agencies pay

   a. the most for freelance work.
   b. the least for freelance work.
   c. on a semiannual basis for freelance work.
   d. on a case-by-case basis.

6. For constant cash flow, a good strategy is to

   a. be willing to work as many as twenty hours a day.
   b. work a little in each of the markets while specializing in one.
   c. diversify your work rather than specialize in one market.
   d. have a full-time job and work as a freelancer part time only.

7. The most common type of usage rights is

   a. one-time usage.
   b. single usage.
   c. first-time usage.
   d. none of the above

8. If unlimited rights are given,

   a. you have locked yourself into a long-term low-pay agreement.
   b. higher pay should be negotiated.
   c. the publisher controls all aspects of your product.
   d. chalk it up to a learning experience, and never give unlimited rights again.

9. The most restrictive rights contract is

   a. unlimited rights.
   b. exclusive rights.
   c. one-time usage rights.
   d. none of the above

10. With articles,

   a. some magazines pay a flat rate, while others pay by the word.
   b. magazines pay only a flat rate.
   c. magazines pay only by the word.
   d. magazines are the hardest publication to break into.

11. A query letter is

   a. your last link to breaking into the freelance business.

b. your first link to breaking into the freelance business.

c. the slowest way to get an editor's attention in order to break into the freelance business.

d. the fastest way to get the publisher's attention in order to break into the freelance business.

12. A query letter must

a. be well written, to the point, and clearly sell your angle.

b. be short and written in block form.

c. politely ask for work.

d. solicit work.

13. As far as deadlines go,

a. make sure you meet most of them.

b. wait until the last day to send in your content.

c. meet them as early as possible.

d. do not concern yourself with them.

14. If an editor calls, writes, or e-mails you,

a. wait at least one working day to respond.

b. respond as quickly as possible.

c. respond as time permits.

d. a response may or may not be in order.

15. Make a habit of

a. contacting editors in the middle of an assignment.

b. getting to work on time.

c. getting to work early and staying late.

d. waiting until your article or photo shoot is finished to contact editors for fear of taking up too much of their time.

16. Knowing the ins and outs of a contract allows you to

a. be a strong negotiator.

b. respectfully decline the offer if the terms do not meet your expectations.

c. get out of a contract if you decide not to do the work.

d. be able to word them to be in your favor.

17. One of the hardest yet most productive things you can learn is to

a. accept lower pay.

b. accept constructive criticism.

c. accept responsibility for your actions.

d. accept assignments that stretch your journalistic abilities.

18. If an editor disagrees with content you know to be factually correct,
    a. argue your point until he or she understands the concept.
    b. be willing to pull your project and pitch the idea to a rival publication.
    c. respectfully disagree and state your case for the disagreement.
    d. stand your ground even if it means stepping on toes to do so.

19. Making a good living as a freelancer means you have to
    a. lower your standard of living.
    b. be willing to go where the news is.
    c. be prolific in producing material.
    d. lower your standard of living until your reputation gets to be known in the industry.

20. Competition
    a. can drive you to become better if you are willing to bend the rules.
    b. can drive you to become better if you are willing to face challenges.
    c. can drive you to become tougher skinned.
    d. can drive you crazy.

Fill in the Blank

angle	capable	size
background	group	specific
clients	wrap	suggestions
closed	angles	terms
core	purchasing	them
cover	step	usage
exactly	proposals	consistently
exclusive	information	writer's
first	masthead	photos
focus	advertising	belongs
guidelines	newspapers	republish
indefinite	abilities	develop
Internet	two	semiregular
magazines	devices	number
negotiate	article	proposed
one	core	length
personal	professional	editorial
photograph	prospect	table
pinnacle	protocol	leased
positive	publication	reproduce
exclusive	publish	agents
information	regular	publish
provide	represent	

1. As a freelance journalist, you may work for several _____ to whom you _____ material on either a _____ or _____-_____ basis.

2. You are free to _____ pay scales and contracts based on your _____.

3. For freelancers, at least four distinct types of markets exist; they are _____, _____, _____, and _____ markets.

4. Magazines are the _____ for many in the freelance business.

5. The key to making a good living as a freelancer is to find a _____ _____ of magazines that _____ your work and are willing to pay you fairly for what you do.

6. Many ad agencies work with _____, who often _____ a _____ of writers or photographers.

7. Working for a _____ does not mean that the work you are producing _____ to _____.

8. The work you produce is _____ for a _____ _____ of time and under specified _____ terms.

9. With unlimited rights, the users of the content can _____ the _____ or the _____ more than once.

10. _____ rights mean that the entity _____ the rights has the _____ right to _____ and _____ the work as they see fit for an _____ period.

11. The _____ is the _____ _____ in contacting a magazine about the _____ of freelance work.

12. In the _____ and photographer's _____, you will find specific _____ on the editorial _____ of the publication, the process by which to submit _____, and the correct _____ for submitting photos and articles.

13. Some magazines pay only one rate for one-time usage of _____ regardless of the _____ that they appear in the magazine.

14. The first paragraph of a query letter should state your _____ topic in _____ or _____ sentences.

15. The second and possibly third paragraph of a query letter should be spent providing _____ _____ on the subject.

16. The third or possibly fourth paragraph of a query letter should spell out _____ the _____ by which you will _____ the subject.

17. The last two paragraphs should _____ up the letter on a _____ note.

18. Knowing the ins and outs of a contract is important, because once

you agree to a set of _____, you cannot go back to the _____ for more pay.

19. An editor's _____ for making a piece of work read better or fit their _____ style better is not a _____ jab.

20. The gap between professional and consumer electronic _____ has _____ considerably. As a result, virtually anyone is _____ of producing _____ Web sites, editorial copy, and artwork.

21. The key to beating the competition _____ is to produce _____ work and _____ fresh ideas with new _____.

ENHANCEMENT ACTIVITIES

1. Using the information found on a magazine's masthead, write to the editor and request a copy of the magazine's writers' and photographers' guidelines.

2. Repeat Activity 1 for multiple magazines, and compare the pay rates offered by the various magazines.

3. Search for freelance photographers on the Internet. Once you locate a photographer, interview the person about being a freelance photographer.

4. Following the format in the chapter, write a query letter.

WEB LINKS

To find out more about the topics discussed in this chapter, visit these Web sites:

What are copyrights and patents?
<http://www.howstuffworks.com/>

Marketing News
<http://www.womeninphotography.org/>

Do's and Don'ts: How to Write the Perfect Query Letter
<http://www.eclectics.com/>
   (Search term: query letter)

The Quest for a Winning Query Letter
<http://www.wga.org/>
   (Search term: query letter)

How to Meet Editors Without Ruining Your Career
<http://w3.one.net/>
  (Search term: meeting)

Alternatively, do your own search at <http://www.alltheweb.com>
  (Search terms: freelance photographer, one-time usage, query letter,
    photographic promotion, magazine editors)

BIBLIOGRAPHY

Writer's Digest Books. (1999). *1999 Photographer's Market*. Cincinnati:
  Author.
Pickerell, Jim, and DeFrank, Cheryl Pickerell. (1997). *Negotiating Stock Photo
  Prices* (4th ed.). Rockville, MD: Author.

CAREERS IN AGRICULTURAL COMMUNICATIONS

OBJECTIVES

After completing this section, you should be able to:

- identify jobs in the agricultural communications field.

- understand the educational requirements of various jobs in the agricultural communications field.

- cite the pay scales of various jobs in the agricultural communications field.

- locate universities that offer an agricultural communications degree.

OVERVIEW

The job prospects for those wishing to communicate in the agriculture industry are numerous. Although it is not rare that you will find a job title that is named agriculture communication specialist, most of the jobs available to agricultural communicators fall under a variety of names.

Agricultural communicators can be camera operators, photographers, writers, editors, design artists, or any number of things. The common denominator is that people in these fields often use their knowledge of agriculture to advance the facts about the industry.

Those wishing to major in and start careers in agricultural communications may find jobs in the advertising, education, editorial, corporate, or industrial arenas of work. As the U.S. Department of Labor (USDOL) points out, the need for people who can communicate in ever-growing informational outlets is growing. Those seeking jobs in agricultural communications should fare well in the marketplace.

AN OVERVIEW OF COMMON AGRICULTURAL COMMUNICATIONS JOBS

Each year, the U.S. Department of Labor, surveys the pay, demand, and educational requirements of every type of job in the United States. Following are excerpts from the USDOL Web site under the Occupational Outlook Handbook, 2002–2003 edition. The main page address is: <http://stats.bls .gov/oco/home.htm>, on jobs that agricultural communicators might hold.

AGRICULTURAL AND FOOD SCIENTISTS

Significant Points

- A large proportion, about 41 percent, of salaried agricultural and food scientists works for Federal, State, and local governments.
- A bachelor's degree in agricultural science is sufficient for some jobs in applied research; a master's or doctoral degree is required for basic research.

Nature of the Work

The work of agricultural and food scientists plays an important part in maintaining the Nation's food supply by ensuring agricultural productivity and the safety of the food supply. Agricultural scientists study farm crops and animals, and develop ways of improving their quantity and quality. They look for ways to improve crop yield with less labor, control pests and weeds more safely and effectively, and conserve soil and water. They research methods of converting raw agricultural commodities into attractive and healthy food products for consumers.

Agricultural science is closely related to biological science, and agricultural scientists use the principles of biology, chemistry, physics, mathematics, and other sciences to solve problems in agriculture. They often work with biological scientists on basic biological research and on applying to agriculture the advances in knowledge brought about by biotechnology.

Many agricultural scientists work in basic or applied research and development. Others manage or administer research and development programs, or manage marketing or production operations in companies that produce food products or agricultural chemicals, supplies, and machinery. Some agricultural scientists are consultants to business firms, private clients, or government.

Depending on the agricultural or food scientist's area of specialization, the nature of the work performed varies.

Food scientists and technologists usually work in the food processing industry, universities, or the Federal Government, and help meet consumer demand for food products that are healthful, safe, palatable, and convenient. To do this, they use their knowledge of chemistry, microbiology, and other sciences to develop new or better ways of preserving, processing, packaging, storing, and delivering foods. Some food scientists engage in basic research, discovering new food sources; analyzing food content to determine levels of vitamins, fat, sugar, or protein; or searching for substitutes for harmful or undesirable additives, such as nitrites.

They also develop ways to process, preserve, package, or store food according to industry and government regulations.

Others enforce government regulations, inspecting food-processing areas and ensuring that sanitation, safety, quality, and waste management standards are met. Food technologists generally work in product development, applying the findings from food science research to the selection, preservation, processing, packaging, distribution, and use of safe, nutritious, and wholesome food.

Agronomy, crop science, entomology, and plant breeding are included in plant science.

Scientists in these disciplines study plants and their growth in soils, helping producers of food, feed, and fiber crops to continue to feed a growing population while conserving natural resources and maintaining the environment. Agronomists and crop scientists not only help increase productivity, but also study ways to improve the nutritional value of crops and the quality of seed. Some crop scientists study the breeding, physiology, and management of crops and use genetic engineering to develop crops resistant to pests and drought. Entomologists conduct research to develop new technologies to control or eliminate pests in infested areas and to prevent the spread of harmful pests to new areas, as well as technologies that are compatible with the environment. They also conduct research or engage in oversight activities aimed at halting the spread of insect-borne disease.

Soil scientists study the chemical, physical, biological, and mineralogical composition of soils as they relate to plant or crop growth. They also study the responses of various soil types to fertilizers, tillage practices, and crop rotation. Many soil scientists who work for the Federal Government conduct soil surveys, classifying and mapping soils. They provide information and recommendations to farmers and other landowners regarding the best use of land, plant growth, and methods to avoid or correct problems such as erosion. They may also consult with engineers and other technical personnel working on construction projects about the effects of, and solutions to, soil problems. Because soil science is closely related to environmental science, persons trained in soil science also apply their knowledge to ensure environmental quality and effective land use.

Animal scientists work to develop better, more efficient ways of producing and processing meat, poultry, eggs, and milk. Dairy scientists, poultry scientists, animal breeders, and other related scientists study the genetics, nutrition, reproduction, growth, and development of domestic farm animals. Some animal scientists inspect and grade livestock food products, purchase livestock, or work in technical sales or marketing.

As extension agents or consultants, animal scientists advise agricultural producers on how to upgrade animal housing facilities properly, lower mortality rates, handle waste matter, or increase production of animal products, such as milk or eggs.

Working Conditions

Agricultural scientists involved in management or basic research tend to work regular hours in offices and laboratories. The work environment for those engaged in applied research or product development varies, depending on the discipline of agricultural science and on the type of employer. For example, food scientists in private industry may work in test kitchens while investigating new processing techniques. Animal scientists working for Federal, State, or university research stations may spend part of their time at dairies, farrowing houses, feedlots, or farm animal facilities or outdoors conducting research associated with livestock. Soil and crop scientists also spend time outdoors conducting research on farms and agricultural research stations.

Entomologists work in laboratories, insectories, or agricultural research stations, and may also spend time outdoors studying or collecting insects in their natural habitat.

Employment

Agricultural and food scientists held about 17,000 jobs in 2000. In addition, several thousand persons held agricultural science faculty positions in colleges and universities.

About 41 percent of all non-faculty salaried agricultural and food scientists work for Federal, State, or local governments. Nearly 2 out of 3 worked for the Federal Government in 2000, mostly in the Department of Agriculture. In addition, large numbers worked for State governments at State agricultural colleges or agricultural research stations. Some worked for agricultural service companies; others worked for commercial research and development laboratories, seed companies, pharmaceutical companies, wholesale distributors, and food products companies.

About 4,000 agricultural scientists were self-employed in 2000, mainly as consultants.

Training, Other Qualifications, and Advancement

Training requirements for agricultural scientists depend on their specialty and on the type of work they perform.

A bachelor's degree in agricultural science is sufficient for some jobs in applied research or for assisting in basic research, but a master's or doctoral degree is required for basic research. A Ph.D. in agricultural science usually is needed for college teaching and for advancement to administrative research positions. Degrees in related sciences such as biology, chemistry, or physics or in related engineering specialties also may qualify persons for some agricultural science jobs.

All States have a land-grant college that offers agricultural science degrees. Many other colleges and universities also offer agricultural science degrees or some agricultural science courses. However, not every school offers all specialties. A typical undergraduate agricultural science curriculum includes communications, economics, business, and physical and life sciences courses, in addition to a wide variety of technical agricultural science courses. For prospective animal scientists, these technical agricultural science courses might include animal breeding, reproductive physiology, nutrition, and meats and muscle biology.

Students preparing as food scientists take courses such as food chemistry, food analysis, food microbiology, food engineering, and food processing operations. Those preparing as crop or soil scientists take courses in plant pathology, soil chemistry, entomology, plant physiology, and biochemistry, among others.

Advanced degree programs include classroom and fieldwork, laboratory research, and a thesis or dissertation based on independent research.

Agricultural and food scientists should be able to work independently or as part of a team and be able to communicate clearly and concisely, both orally and in writing. Most of these scientists also need an understanding of basic business principles, and the ability to apply basic statistical techniques. Employers increasingly prefer job applicants who are able to apply computer skills to determine solutions to problems, to collect and analyze data, and for the control of processes.

The American Society of Agronomy offers certification programs in crops, agronomy, crop advising, soils, horticulture, plant pathology, and weed science. To become certified, applicants must pass designated examinations and meet certain standards with respect to education and professional work experience.

Agricultural scientists who have advanced degrees usually begin in research or teaching. With experience, they may advance to jobs such as supervisors of research programs or managers of other agriculture-related activities.

Job Outlook

Employment of agricultural scientists is expected to grow more slowly than the average for all occupations through 2010. Additionally, the need to replace agricultural and food scientists who retire or otherwise leave the occupation permanently will account for many more job openings than will projected growth, particularly in academia.

Past agricultural research has resulted in the development of higher yielding crops, crops with better resistance to pests and plant pathogens, and chemically based fertilizers and pesticides. Further research is necessary as insects and diseases continue to adapt to pesticides, and as soil fertility and water quality continue to need improvement. Agricultural scientists are using new avenues of research in biotechnology to develop plants and food crops that require less fertilizer, fewer pesticides and herbicides, and even less water for growth.

Agricultural scientists will be needed to balance increased agricultural output with protection and preservation of soil, water, and ecosystems. They will increasingly encourage the practice of "sustainable agriculture" by developing and implementing plans to manage pests, crops, soil fertility and erosion, and animal waste in ways that reduce the use of harmful chemicals and do little damage to the natural environment. In addition, an expanding population and an increasing public focus on diet, health, and food safety will result in job opportunities for food scientists and technologists.

Graduates with advanced degrees will be in the best position to enter jobs as agricultural scientists.

Bachelor's degree holders can work in some applied research and product development positions, but usually only in certain sub fields, such as food science and technology. In addition, the Federal Government hires bachelor's degree holders to work as soil scientists. Despite the more limited opportunities for those with only a bachelor's degree to obtain jobs as agricultural scientists, a bachelor's degree in agricultural science is useful for managerial jobs in businesses that deal with ranchers and farmers, such as feed, fertilizer, seed, and farm equipment manufacturers; retailers or wholesalers; and farm credit institutions. Four-year degrees also may help persons enter occupations such as farmer, or farm or ranch manager; cooperative extension service agent; agricultural products inspector; or purchasing or sales agent for agricultural commodity or farm supply companies.

Earnings

Median annual earnings of agricultural and food scientists were $52,160 in 2000. The middle 50 percent earned between $40,720 and $66,370. The lowest 10 percent earned less than $31,910, and the highest 10 percent earned more than $83,740.

Average Federal salaries for employees in non-supervisory, supervisory and managerial positions in certain agricultural science specialties in 2001 were as follows: Animal science, $76,582; agronomy, $62,311; soil science, $58,878; horticulture, $59,472; and entomology, $70,133.

According to the National Association of Colleges and Employers, beginning salary offers in 2001 for graduates with a bachelor's degree in animal science averaged $28,031 a year.

Related Occupations

The work of agricultural scientists is closely related to that of biologists and other natural scientists, such as chemists, conservation scientists, and foresters. It is also related to managers of agricultural production, such as farmers, ranchers, agricultural managers. Certain specialties of agricultural science also are related to other occupations. For example, the work of animal scientists is related to that of veterinarians and horticulturists perform duties similar to those of landscape architects.

Sources of Additional Information

Information on careers in agricultural science is available from:

- American Society of Agronomy, Crop Science Society of America, Soil Science Society of America, 677 S. Segoe Rd., Madison, WI 53711-1086.
- Food and Agricultural Careers for Tomorrow, Purdue University, 1140 Agricultural Administration Bldg., West Lafayette, IN 47907-1140.

For information on careers in food technology, write to:

- Institute of Food Technologists, Suite 300, 221 N. LaSalle St., Chicago IL 60601-1291.

Information on acquiring a job as an agricultural scientist with the Federal Government is available from the Office of Personnel Management through a telephone-based system. Consult your telephone directory under U.S. Government for a local number or call (912) 757-3000; Federal Relay Service: (800) 877-8339. The first number is not toll free, and charges may result. Information also is available from the Internet site: <http://www .usajobs.opm.gov>.

ANNOUNCERS

Significant Points

- Competition for announcer jobs will continue to be keen.
- Jobs at small stations usually have low pay, but offer the best opportunities for beginners.
- Related work experience at a campus radio station or as an intern at a commercial station can be helpful in breaking into the occupation.

Nature of the Work

Announcers in radio and television perform a variety of tasks on and off the air. They announce station program information such as program schedules and station breaks for commercials or public service information, and they introduce and close programs. Announcers read prepared scripts or ad-lib commentary on the air when presenting news, sports, weather, time, and commercials. If a written script is required, they may do the research and writing. Announcers also interview guests and moderate panels or discussions. Some provide commentary for the audience during sporting events, parades, and other events. Announcers are often well known to radio and television audiences and may make promotional appearances and remote broadcasts for their stations.

Newscasters or anchors work at large stations and specialize in news, sports, or weather. Show hosts may specialize in a certain area of interest such as politics, personal finance, sports, or health. They contribute to the preparation of the program content; interview guests; and discuss issues with viewers, listeners, or an in-studio audience.

Announcers at smaller stations may cover all of these areas and tend to have more off-air duties as well. They may operate the control board, monitor the transmitter, sell commercial time to advertisers, keep a log of the station's daily programming, and do production work. Consolidation and automation make it possible for announcers to do some work previously performed by broadcast technicians.

Announcers use the control board to broadcast programming, commercials, and public service announcements according to schedule. Public radio and television announcers are involved with station fundraising efforts.

Although most announcers are employed in radio and television broadcasting, some are employed in the cable television or motion picture production industries. Other announcers may use a public address system to provide information to the audience at sporting and other events.

Working Conditions

Announcers usually work in well-lighted, air-conditioned, soundproof studios. The broadcast day is long for radio and TV stations—some are on the air 24 hours a day—so announcers can expect to work unusual hours.

Many present early morning shows, when most people are getting ready for work or commuting, while others do late night programs.

Announcers often work within tight schedule constraints, which can be physically and mentally stressful. For many announcers, the intangible rewards—creative work, many personal contacts, and the satisfaction of becoming widely known—far outweigh the disadvantages of irregular and often unpredictable hours, work pressures, and disrupted personal lives.

Employment

Announcers held about 71,000 jobs in 2000. Nearly all were staff announcers employed in radio and television broadcasting, but some were freelance announcers who sold their services for individual assignments to networks and stations, or to advertising agencies and other independent producers. Many announcing jobs are part time.

Training, Other Qualifications, and Advancement

Entry into this occupation is highly competitive. Formal training in broadcasting from a college or technical school (private broadcasting school) is valuable. Station officials pay particular attention to taped auditions that show an applicant's delivery and—in television—appearance and style on commercials, news, and interviews.

Those hired by television stations usually start out as production assistants, researchers, or reporters and are given a chance to move into announcing if they show an aptitude for "on-air" work. Newcomers to TV broadcasting also may begin as news camera operators. . . . A beginner's chance of landing an on-air job is remote, except possibly for a small radio station. In radio, newcomers usually start out taping interviews and operating equipment.

Announcers usually begin at a station in a small community and, if qualified, may move to a better paying job in a large city. They also may advance by hosting a regular program as a disc jockey, sportscaster, or other specialist.Competition is particularly intense for employment by networks, and employers look for college graduates with at least several years of successful announcing experience.

Announcers must have a pleasant and well-controlled voice, good timing, excellent pronunciation, and must know correct grammar usage. Television announcers need a neat, pleasing appearance as well. Announcers also must be computer-literate because programming is created and edited by computer.

In addition, they should be able to ad-lib all or part of a show and to work under tight deadlines. The most successful announcers attract a large audience by combining a pleasing personality and voice with an appealing style.

High school and college courses in English, public speaking, drama, foreign languages, and computer science are valuable, and hobbies such as sports and music are additional assets. Students may gain valuable experience at campus radio or TV facilities and at commercial stations while serving as interns. Paid or unpaid internships provide students with hands-on training and the chance to establish contacts in the industry. Unpaid interns often receive college credit and are allowed to observe and assist station employees. Although the Fair Labor Standards Act limits the work unpaid interns may perform in a station, unpaid internships are the rule; sometimes they lead to paid internships. Paid internships are valuable because interns do work ordinarily done by regular employees and may even go on the air.

Persons considering enrolling in a broadcasting school should contact personnel managers of radio and television stations as well as broadcasting trade organizations to determine the school's reputation for producing suitably trained candidates.

Job Outlook

Competition for jobs as announcers will be keen because the broadcasting field attracts many more job seekers than there are jobs. Small radio stations are more inclined to hire beginners, but the pay is low. Interns usually receive preference for available positions. Because competition for ratings is so intense in major metropolitan areas, large stations will continue to seek announcers who have proven that they can attract and retain a large audience.

Announcers who are knowledgeable in business, consumer, and health news may have an advantage over others. While specialization is more common at large stations and the networks, many small stations also encourage it.

Employment of announcers is expected to decline through 2010 due to the lack of growth of new radio and television stations. Openings in this relatively small field also will arise from the need to replace those who transfer to other kinds of work or leave the labor force. Some announcers leave the field because they cannot advance to better paying jobs. Changes in station ownership, format, and ratings frequently cause periods of unemployment for many announcers.

Increasing consolidation of radio and television stations, new technology, and the growth of alternative media sources will contribute to the expected decline in employment of announcers. Consolidation in broadcasting may lead to increased use of syndicated programming and programs originating outside a station's viewing or listening area. Digital technology will increase the productivity of announcers, reducing the time spent on off-air technical and production work. In addition, all traditional media, including radio and television, may suffer losses in audience as the American public increases its use of personal computers.

Earnings

Salaries in broadcasting vary widely but in general are relatively low, except for announcers who work for large stations in major markets or for networks. Earnings are higher in television than in radio and higher in commercial than in public broadcasting.

Median hourly earnings of announcers in 2000 were $9.52. The middle 50 percent earned between $6.84 and $14.28. The lowest 10 percent earned less than $5.94, and the highest 10 percent earned more than $24.35.

Median hourly earnings of announcers in 2000 were $9.54 in the radio and television broadcasting industry.

Related Occupations

The success of announcers depends upon how well they communicate. Others who must be skilled at oral communication include news analysts, reporters, and correspondents; interpreters and translators; sales and related occupations; public relations specialists; and teachers. Many announcers also must entertain their audience, so their work is similar to other entertainment-related occupations such as actors, directors, and producers; dancers and choreographers; and musicians, singers, and related workers.

Sources of Additional Information

General information on the broadcasting industry is available from:

- National Association of Broadcasters, 1771 N St. NW., Washington, DC 20036. Internet: <http://www.nab.org>.

ARTISTS AND RELATED WORKERS

- More than half are self-employed—about seven times the proportion in all professional and related occupations.
- Artists usually develop their skills through a bachelor's degree program or other postsecondary training in art or design.
- Keen competition is expected for both salaried jobs and freelance work, because many talented people are attracted to the visual arts.

Nature of the Work

Artists create art to communicate ideas, thoughts, or feelings. They use a variety of methods—painting, sculpting, or illustration—and an assortment of materials, including oils, watercolors, acrylics, pastels, pencils, pen and ink, plaster, clay, and computers. Artists' works may be realistic, stylized, or abstract and may depict objects, people, nature, or events.

Artists generally fall into one of three categories. Art directors formulate design concepts and presentation approaches for visual communications media. Fine artists, including painters, sculptors, and illustrators create original artwork using a variety of media and techniques. Multimedia artists and animators create special effects, animation, or other visual images using film, video, computers or other electronic media.

Art directors develop design concepts and review the material that is to appear in periodicals, newspapers, and other printed or digital media. They decide how best to present the information visually, so it is eye-catching, appealing, and organized. They decide which photographs or artwork to use and oversee the layout design and production of the printed material. They may direct workers engaged in artwork, layout design, and copy writing.

Illustrators typically create pictures for books, magazines, and other publications; and commercial products, such as textiles, wrapping paper, stationary, greeting cards and calendars. Increasingly, illustrators work in digital format, preparing work directly on a computer.

Medical and scientific illustrators combine drawing skills with knowledge of the biological sciences. Medical illustrators draw illustrations of human anatomy and surgical procedures. Scientific illustrators draw illustrations of animals and plants.

Multi-media artists and animators work primarily in computer and data processing services, advertising, and the motion picture and television industries. They draw by hand and use computers to create the large series of pictures that form the animated images or special effects seen in movies, television programs, and computer games. Some draw storyboards for television commercials, movies, and animated features. Storyboards present television commercials in a series of scenes similar to a comic strip and allow an advertising agency to evaluate proposed commercials with the company doing the advertising. Storyboards also serve as guides to placing actors and cameras and to other details during the production of commercials.

Working Conditions

Most artists work in fine or commercial art studios located in office buildings, or in private studios in their homes. Some fine artists share studio space, where they also may exhibit their work.

Artists employed by publishing companies, advertising agencies, and design firms generally work a standard 40-hour week. During busy periods, they may work overtime to meet deadlines. Self-employed artists can set their own hours, but may spend much time and effort selling their artwork to potential customers or clients and building a reputation.

Employment

Artists held about 147,000 jobs in 2000. More than half were self-employed. Of the artists who were not self-employed, many worked in motion picture, television, computer software, printing, publishing, and public relations firms. Some self-employed artists offer their services to advertising agencies, design firms, publishing houses, and other businesses.

Training, Other Qualifications, and Advancement

Training requirements for artists vary by specialty. Although formal training is not strictly necessary for fine artists, it is very difficult to become skilled enough to make a living without some training. Many colleges and universities offer degree programs leading to the Bachelor in Fine Arts (BFA) and Master in Fine Arts (MFA) degrees. Coursework usually includes core subjects, such as English, social science, and natural science, in addition to art history and studio art.

Independent schools of art and design also offer postsecondary studio training in the fine arts leading to an Associate in Art (AA) or Bachelor in Fine Arts (BFA) degree. Typically, these programs focus more intensively on studio work than the academic programs in a university setting.

Formal educational programs in art also provide training in computer techniques. Computers are used widely in the visual arts, and knowledge and training in them are critical for many jobs in these fields.

Illustrators learn drawing and sketching skills through training in art programs and extensive practice. Most employers prefer candidates with a bachelor's degree; however, some illustrators are contracted based on their portfolios of past work.

Medical illustrators must have both a demonstrated artistic ability and a detailed knowledge of living organisms, surgical and medical procedures, and human and animal anatomy. A 4-year bachelor's degree combining art and premedical courses usually is preferred, followed by a master's degree in medical illustration.

This degree is offered in only five accredited schools in the United States.

Evidence of appropriate talent and skill, displayed in an artist's portfolio, is an important factor used by art directors, clients, and others in deciding whether to hire or contract out work.

The portfolio is a collection of hand-made, computer-generated, photographic, or printed samples of the artist's best work. Assembling a successful portfolio requires skills usually developed in a bachelor's degree program or other postsecondary training in art or visual communications. Internships also provide excellent opportunities for artists to develop and enhance their portfolios.

Artists hired by advertising agencies often start with relatively routine work. While doing this work, however, they may observe and practice their skills on the side. Many artists freelance on a part-time basis while continuing to hold a full-time job until they are established. Others freelance part-time while still in school, to develop experience and to build a portfolio of published work.

Freelance artists try to develop a set of clients who regularly contract for work. Some freelance artists are widely recognized for their skill in specialties such as magazine or children's book illustration. These artists may earn high incomes and can choose the type of work they do.

Fine artists advance professionally as their work circulates and as they establish a reputation for a particular style. Many of the most successful artists continually develop new ideas, and their work often evolves over time.

Job Outlook

Employment of artists and related workers is expected to grow as fast as the average for all occupations through the year 2010. Because the arts attract many talented people with creative ability, the number of aspiring artists continues to grow. Consequently, competition for both salaried jobs and freelance work in some areas is expected to be keen.

Fine artists mostly work on a freelance, or commission, basis and may find it difficult to earn a living solely by selling their artwork. Only the most successful fine artists receive major commissions for their work.

Population growth, rising incomes, and growth in the number of people who appreciate the fine arts will contribute to the demand for fine artists. Talented fine artists, who have developed a mastery of artistic techniques and skills, including computer skills, will have the best job prospects.

The need for artists to illustrate and animate materials for magazines, journals, and other printed or electronic media will spur demand for illustrators and animators of all types. Growth in the entertainment industry, including cable and other pay television broadcasting and motion picture production and distribution, will provide new job opportunities for illustrators, cartoonists, and animators. Competition for most jobs, however, will be strong, because job opportunities are relatively few and the number of people interested in these positions usually exceeds the number of available openings. Employers should be able to choose from among the most qualified candidates.

Earnings

Median annual earnings of salaried art directors were $56,880 in 2000. The middle 50 percent earned between $41,290 and $80,350. The lowest 10 percent earned less than $30,130, and the highest 10 percent earned more than $109,440. Median annual earnings were $63,510 in advertising, the industry employing the largest numbers of salaried art directors.

Median annual earnings of salaried fine artists, including painters, sculptors, and illustrators were $31,190 in 2000. The middle 50 percent earned between $20,460 and $42,720. The lowest 10 percent earned less than $14,690, and the highest 10 percent earned more than $58,580.

Median annual earnings of salaried multi-media artists and animators were $41,130 in 2000. The middle 50 percent earned between $30,700 and $54,040. The lowest 10 percent earned less than $23,740, and the highest 10 percent earned more than $70,560. Median annual earnings were $44,290 in computer and data processing services, the industry employing the largest numbers of salaried multi-media artists and animators.

Earnings for self-employed artists vary widely. Some charge only a nominal fee while they gain experience and build a reputation for their work. Others, such as well-established freelance fine artists and illustrators, can earn more than salaried artists can. Many, however, find it difficult to rely solely on income earned from selling paintings or other works of art. Like other self-employed workers, freelance artists must provide their own benefits.

Sources of Additional Information

For general information about art and design and a list of accredited college-level programs, contact:

- The National Association of Schools of Art and Design, 11250 Roger Bacon Dr., Suite 21, Reston, VA 20190. Internet: <http://www.arts-accredit.org> (Search word: nasad)

BROADCAST AND SOUND ENGINEERING TECHNICIANS AND RADIO OPERATORS

Significant Points

- Job applicants will face strong competition for the better paying jobs at radio and television stations serving large cities.
- Television stations employ, on average, many more technicians than do radio stations.
- Evening, weekend, and holiday work is common.

Nature of the Work

Broadcast and sound engineering technicians install, test, repair, set up, and operate the electronic equipment used to record and transmit radio and television programs, cable programs, and motion pictures. They work with television cameras, microphones, tape recorders, lighting, sound effects, transmitters, antennas, and other equipment. Some broadcast and sound engineering technicians produce movie soundtracks in motion picture production studios, control the sound of live events, such as concerts, or record music in a recording studio.

In the control room of a radio or television-broadcasting studio, these technicians operate equipment that regulates the signal strength, clarity, and range of sounds and colors of recordings or broadcasts. They also operate control panels to select the source of the material. Technicians may switch from one camera or studio to another, from film to live programming, or from network to local programming. By means of hand signals and, in television, telephone headsets, they give technical directions to other studio personnel.

Audio and video equipment operators operate specialized electronic equipment to record stage productions, live programs or events, and studio recordings. They edit and reproduce tapes for compact discs, records and cassettes, for radio and television broadcasting and for motion picture productions. The duties of audio and video equipment operators can be divided into two categories: technical and production activities used in the production of sound and picture images for film or videotape from set design to camera operation and post production activities where raw images are transformed to a final print or tape.

Radio operators mainly receive and transmit communications using a variety of tools. They are also responsible for repairing equipment using such devices as electronic testing equipment, hand tools, and power tools. These help to maintain communication systems in an operative condition.

Broadcast and sound engineering technicians and radio operators perform a variety of duties in small stations.

In large stations and at the networks, technicians are more specialized, although job assignments may change from day to day. The terms "operator," "engineer," and "technician" often are used interchangeably to describe these jobs. Transmitter operators monitor and log outgoing signals and operate transmitters. Maintenance technicians set up, adjust, service, and repair electronic broadcasting equipment. Audio control engineers regulate volume and sound quality of television broadcasts, while video control engineers regulate their fidelity, brightness, and contrast. Recording engineers operate and maintain video and sound recording equipment. They may operate equipment designed to produce special effects, such as the illusions of a bolt of lightning or a police siren. Sound mixers or re-recording mixers produce the sound track of a movie, television, or radio program.

After filming or recording, they may use a process called dubbing to insert sounds. Field technicians set up and operate broadcasting portable field transmission equipment outside the studio.

Television news coverage requires so much electronic equipment, and the technology is changing so rapidly, that many stations assign technicians exclusively to news.

Chief engineers, transmission engineers, and broadcast field supervisors supervise the technicians who operate and maintain broadcasting equipment.

Working Conditions

Broadcast, sound engineering, audio and video equipment technicians, and radio operators generally work indoors in pleasant surroundings. However, those who broadcast news and other programs from locations outside the studio may work outdoors in all types of weather. Technicians doing maintenance may climb poles or antenna towers, while those setting up equipment do heavy lifting.

Technicians in large stations and the networks usually work a 40-hour week under great pressure to meet broadcast deadlines, but may occasionally work overtime. Technicians in small stations routinely work more than 40 hours a week. Evening, weekend, and holiday work is usual, because most stations are on the air 18 to 24 hours a day, 7 days a week.

Those who work on motion pictures may be on a tight schedule to finish according to contract agreements.

Employment

Broadcast and sound engineering technicians and radio operators held about 87,000 jobs in 2000. Their employment was distributed among the following detailed occupations:

Audio and video equipment technicians	37,000
Broadcast technicians	36,000
Sound engineering technicians	11,000
Radio operators	2,900

About 1 out of 3 worked in radio and television broadcasting. Almost 15 percent worked in the motion picture industry. About 4 percent worked for cable and other pay-television services. A few were self-employed. Television stations employ, on average, many more technicians than do radio stations. Some technicians are employed in other industries, producing employee communications, sales, and training programs.

Technician jobs in television are located in virtually all cities, whereas jobs in radio are also found in many small towns. The highest paying and most specialized jobs are concentrated in New York City, Los Angeles, Chicago, and Washington, DC—the originating centers for most network programs. Motion picture production jobs are concentrated in Los Angeles and New York City.

Training, Other Qualifications, and Advancement

The best way to prepare for a broadcast and sound engineering technician job is to obtain technical school, community college, or college training in broadcast technology or in engineering or electronics. This is particularly true for those who hope to advance to supervisory positions or jobs at large stations or the networks.

In the motion picture industry people are hired as apprentice editorial assistants and work their way up to more skilled jobs. Employers in the motion picture industry usually hire experienced freelance technicians on a picture-by-picture basis. Reputation and determination are important in getting jobs.

Beginners learn skills on the job from experienced technicians and supervisors. They often begin their careers in small stations and, once experienced, move on to larger ones. Large stations usually only hire technicians with experience. Many employers pay tuition and expenses for courses or seminars to help technicians keep abreast of developments in the field.

Audio and video equipment technicians generally need a high school diploma. Many recent entrants have a community college degree or various other forms of post-secondary degrees, although that is not always a requirement. They may substitute on-the-job training for formal education requirements. Experience in a recording studio, as an assistant, is a great way of getting experience and knowledge simultaneously.

Radio operators do not usually require any formal training. This is an entry-level position that generally requires on-the-job training.

The Federal Communications Commission no longer requires the licensing of broadcast technicians, as the Telecommunications Act of 1996 eliminated this licensing requirement. Certification by the Society of Broadcast Engineers is a mark of competence and experience. The certificate is issued to experienced technicians who pass an examination. By offering the Radio Operator and the Television Operator levels of certification, the Society of Broadcast Engineers has filled the void left by the elimination of the FCC license.

Prospective technicians should take high school courses in math, physics, and electronics. Building electronic equipment from hobby kits and operating a "ham," or amateur radio, are good experience, as is work in college radio and television stations.

Broadcast and sound engineering technicians and radio operators must have manual dexterity and an aptitude for working with electrical, electronic, and mechanical systems and equipment.

Experienced technicians can become supervisory technicians or chief engineers. A college degree in engineering is needed to become chief engineer at a large TV station.

Job Outlook

People seeking entry-level jobs as technicians in the field of radio and television broadcasting are expected to face strong competition in major metropolitan areas, where pay generally is higher and the number of qualified job seekers exceed the number of openings. There, stations seek highly experienced personnel. Prospects for entry-level positions generally are better in small cities and towns for beginners with appropriate training.

The overall employment of broadcast and sound engineering technicians and radio operators is expected to grow about as fast as the average for all occupations through the year 2010. An increase in the number of programming hours should require additional technicians. However, employment growth in radio and television broadcasting may be tempered somewhat because of slow growth in the number of new radio and television stations and laborsaving technical advances, such as computer-controlled programming and remote control of transmitters. Technicians who know how to install transmitters will be in demand as television stations replace existing analog transmitters with digital transmitters. Stations will begin broadcasting in both analog and digital formats, eventually switching entirely to digital.

Employment of broadcast and sound engineering technicians is expected to grow about as fast as average through 2010. The advancements in technology will enhance the capabilities of technicians to help produce a higher quality of programming for radio and television. Employment of audio and video equipment technicians also is expected to grow about as fast as average through 2010. Not only will these workers have to set up audio and video equipment, but it will be necessary for them to maintain and repair this machinery. Employment of radio operators, on the other hand, will grow more slowly than other areas in this field of work. Automation will negatively impact these workers as many stations now operate transmitters and control programming remotely.

Employment of broadcast and sound engineering technicians and radio operators in the cable industry should grow rapidly because of new products coming to market, such as cable modems, which deliver high-speed Internet access to personal computers, and digital set-top boxes, which transmit better sound and pictures, allowing cable operators to offer many more channels than in the past. These new products should cause traditional cable subscribers to sign up for additional services.

Employment in the motion picture industry also will grow fast. However, job prospects are expected to remain competitive, because of the large number of people attracted to this relatively small field.

Numerous job openings also will result from the need to replace experienced technicians who leave the occupations. Many leave these occupations for electronic jobs in other areas, such as computer technology or commercial and industrial repair.

Earnings

Television stations usually pay higher salaries than radio stations; commercial broadcasting usually pays more than public broadcasting; and stations in large markets pay more than those in small ones.

Median annual earnings of broadcast technicians in 2000 were $26,950. The middle 50 percent earned between $18,060 and $44,410. The lowest 10 percent earned less than $13,860, and the highest 10 percent earned more than $63,340.

Median annual earnings of sound engineering technicians in 2000 were $39,480. The middle 50 percent earned between $24,730 and $73,720. The lowest 10 percent earned less than $17,560, and the highest 10 percent earned more than $119,400.

Median annual earnings of audio and video equipment technicians in 2000 were $30,310. The middle 50 percent earned between $21,980 and $44,970. The lowest 10 percent earned less than $16,630, and the highest 10 percent earned more than $68,720.

Median annual earnings of radio operators in 2000 were $29,260. The middle 50 percent earned between $23,090 and $39,830. The lowest 10 percent earned less than $17,570, and the highest 10 percent earned more than $54,590.

Related Occupations

Broadcast and sound engineering technicians and radio operators need the electronics training and hand coordination necessary to operate technical equipment, and they generally complete specialized postsecondary programs. Similar occupations include engineering technicians, science technicians, health technologists and technicians, electrical and electronics installers and repairers, and communications equipment operators.

Sources of Additional Information

- For information on careers for broadcast and sound engineering technicians and radio operators, write to:

  National Association of Broadcasters, 1771 N St. NW., Washington, DC 20036. Internet: <http://www.nab.org>

- For information on certification, contact:

  Society of Broadcast Engineers, 9247 North Meridian St., Suite 305, Indianapolis, IN 46260. Internet: <http://www.sbe.org>

- For information on careers in the motion picture and television industry, contact:

  Society of Motion Picture and Television Engineers (SMPTE), 595 West Hartsdale Ave., White Plains, NY 10607. Internet: <http://www.smpte.org>

DESIGNERS

Significant Points

- Three out of 10 designers are self-employed—almost 5 times the proportion for all professional and related occupations.
- Creativity is crucial in all design occupations; most designers need a bachelor's degree, and candidates with a master's degree hold an advantage.
- Keen competition is expected for most jobs, despite projected faster-than-average employment growth, because many talented individuals are attracted to careers as designers.

Nature of the Work

Designers are people with a desire to create. They combine practical knowledge with artistic ability to turn abstract ideas into formal designs for the merchandise we buy, the clothes we wear, the publications we read, and the living and office space we inhabit. Designers usually specialize in a particular area of design, such as automobiles, industrial or medical equipment, or home appliances; clothing and textiles; floral arrangements; publications, logos, signage, or movie or TV credits; interiors of homes or office buildings; merchandise displays; or movie, television, and theater sets.

The first step in developing a new design or altering an existing one is to determine the needs of the client, the ultimate function for which the design is intended, and its appeal to customers. When creating a design, designers often begin by researching the desired design characteristics, such as size, shape, weight, color, materials used, cost, ease of use, fit, and safety.

Designers then prepare sketches—by hand or with the aid of a computer—to illustrate the vision for the design. After consulting with the client, an art or design director, or a product development team, designers create detailed designs using drawings, a structural model, computer simulations, or a full-scale prototype. Many designers increasingly are using computer-aided design (CAD) tools to create and better visualize the final product. Computer models allow greater ease and flexibility in exploring a greater number of design alternatives, thus reducing design costs and cutting the time it takes to deliver a product to market. Industrial designers use computer-aided industrial design (CAID) tools to create designs and machine-readable instructions that communicate with automated production tools.

Designers sometimes supervise assistants who carry out their creations. Designers who run their own businesses also may devote a considerable amount of time to developing new business contacts, reviewing equipment and space needs, and performing administrative tasks, such as reviewing catalogues and ordering samples. Design encompasses a number of different fields. Many designers specialize in a particular area of design, whereas others work in more than one area.

Commercial and industrial designers, including designers of commercial products and equipment, develop countless manufactured products. They combine artistic talent with research on product use, customer needs, marketing, materials, and production methods to create the most functional and appealing design that will be competitive with others in the marketplace. Industrial designers typically concentrate in an area of subspecialization such as kitchen appliances, auto interiors, or plastic-molding machinery.

Graphic designers use a variety of print, electronic, and film media to create designs that meet clients' commercial needs. Using computer software, they develop the overall layout and design of magazines, newspapers, journals, corporate reports, and other publications. They also may produce promotional displays and marketing brochures for products and services, design distinctive company logos for products and businesses, and develop signs and signage systems—called environmental graphics—for business and government. An increasing number of graphic designers develop material to appear on Internet home pages.

Graphic designers also produce the credits that appear before and after television programs and movies.

Working Conditions

Working conditions and places of employment vary. Designers employed by manufacturing establishments, large corporations, or design firms generally work regular hours in well-lighted and comfortable settings. Self-employed designers tend to work longer hours.

Designers who work on a contract, or job, basis frequently adjust their workday to suit their clients' schedules, meeting with them during evening or weekend hours when necessary. Designers may transact business in their own offices or studios or in clients' homes or offices, or they may travel to other locations, such as showrooms, design centers, clients' exhibit sites, and manufacturing facilities. Designers who are paid by the assignment are under pressure to please clients and to find new ones to maintain a constant income. All designers face frustration at times when their designs are rejected or when they cannot be as creative as they wish. With the increased use of computers in the workplace and the advent of Internet websites, more designers conduct business, research design alternatives, and purchase supplies electronically than ever before.

Occasionally, industrial designers may work additional hours to meet deadlines. Similarly, graphic designers usually work regular hours, but may work evenings or weekends to meet production schedules.

Employment

Designers held about 492,000 jobs in 2000. About one-third were self-employed. Employment was distributed as follows:

Graphic designers 190,000

Designers work in a number of different industries, depending on their design specialty. Most industrial designers, for example, work for engineering or architectural consulting firms or for large corporations. Others are self-employed and do freelance work—full time or part time—in addition to a salaried job in another occupation.

Training, Other Qualifications, and Advancement

Creativity is crucial in all design occupations. People in this field must have a strong sense of the esthetic—an eye for color and detail, a sense of balance and proportion, and an appreciation for beauty. Despite the advancement of computer-aided design, sketching ability remains an important advantage in most types of design, especially fashion design. A good portfolio—a collection of examples of a person's best work—often is the deciding factor in getting a job.

A bachelor's degree is required for most entry-level design positions, except for floral design and visual merchandising. Esthetic ability is important for floral design and visual merchandising, but formal preparation typically is not necessary. Many candidates in industrial design pursue a master's degree to better compete for open positions.

Formal training for some design professions also is available in 2- and 3-year professional schools that award certificates or associate degrees in design. Graduates of 2-year programs normally qualify as assistants to designers. The Bachelor of Fine Arts degree is granted at 4-year colleges and universities. The curriculum in these schools includes art and art history, principles of design, designing and sketching, and specialized studies for each of the individual design disciplines, such as garment construction, textiles, mechanical and architectural drawing, computerized design, sculpture, architecture, and basic engineering. A liberal arts education, with courses in merchandising, business administration, marketing, and psychology, along with training in art, is recommended for designers who want to freelance. Additionally, persons with training or experience in architecture qualify for some design occupations, particularly interior design.

Because computer-aided design is increasingly common, many employers expect new designers to be familiar with its use as a design tool. For example, industrial designers extensively use computers in the aerospace, automotive, and electronics industries. Interior designers use computers to create numerous versions of interior space designs—images can be inserted, edited, and replaced easily and without added cost—making it possible for a client to see and choose among several designs.

The National Association of Schools of Art and Design currently accredits about 200 postsecondary institutions with programs in art and design; most of these schools award a degree in art. Some award degrees in industrial, interior, textile, graphic, or fashion design. Many schools do not allow formal entry into a bachelor's degree program until a student has successfully finished a year of basic art and design courses. Applicants may be required to submit sketches and other examples of their artistic ability.

Individuals in the design field must be creative, imaginative, persistent, and able to communicate their ideas in writing, visually, and verbally. Because tastes in style and fashion can change quickly, designers need to be well read, open to new ideas and influences, and quick to react to changing trends. Problem-solving skills and the ability to work independently and under pressure are important traits. People in this field need self-discipline to start projects on their own, to budget their time, and to meet deadlines and production schedules. Good business sense and sales ability also are important, especially for those who freelance or run their own business.

Beginning designers usually receive on-the-job training, and normally need 1 to 3 years of training before they can advance to higher-level positions. Experienced designers in large firms may advance to chief designer, design department head, or other supervisory positions. Some designers become teachers in design schools and colleges and universities. Many faculty members continue to consult privately or operate small design studios to complement their classroom activities. Some experienced designers open their own firms.

Job Outlook

Despite projected faster-than-average employment growth, designers in most fields—with the exception of floral design—are expected to face keen competition for available positions. Many talented individuals are attracted to careers as designers. Individuals with little or no formal education in design, as well as those who lack creativity and perseverance, will find it very difficult to establish and maintain a career in design.

Overall, the employment of designers is expected to grow faster than the average for all occupations through the year 2010. In addition to those that result from employment growth, many job openings will arise from the need to replace designers who leave the field. Increased demand for industrial designers will stem from the continued emphasis on product quality and safety; the demand for new products that are easy and comfortable to use; the development of high-technology products in medicine, transportation, and other fields; and growing global competition among businesses. Demand for graphic designers should increase because of the rapidly increasing demand for Web-based graphics and the expansion of the video entertainment market, including television, movies, videotape, and made-for-Internet outlets.

Earnings

Median annual earnings for commercial and industrial designers were $48,780 in 2000. The middle 50 percent earned between $36,460 and

$64,120. The lowest 10 percent earned less than $27,290, and the highest 10 percent earned more than $77,790.

Median annual earnings for fashion designers were $48,530 in 2000. The middle 50 percent earned between $34,800 and $73,780. The lowest 10 percent earned less than $24,710, and the highest 10 percent earned more than $103,970. Median annual earnings were $52,860 in apparel, piece goods, and notions—the industry employing the largest numbers of fashion designers.

Median annual earnings for floral designers were $18,360 in 2000. The middle 50 percent earned between $14,900 and $22,110. The lowest 10 percent earned less than $12,570, and the highest 10 percent earned more than $27,860. Median annual earnings were $20,160 in grocery stores and $17,760 in miscellaneous retail stores, including florists.

Median annual earnings for graphic designers were $34,570 in 2000. The middle 50 percent earned between $26,560 and $45,130. The lowest 10 percent earned less than $20,480, and the highest 10 percent earned more than $58,400. Median annual earnings in the industries employing the largest numbers of graphic designers were as follows:

Management and public relations	$37,570
Advertising	$37,080
Mailing, reproduction, and stenographic services	$36,130
Commercial printing	$29,730
Newspapers	$28,170

The American Institute of Graphic Arts (AIGA) reported 1999 median earnings for graphic designers with increasing levels of responsibility. Staff-level graphic designers earned $36,000, while senior designers, who may supervise junior staff or have some decision-making authority that reflects their knowledge of graphic design, earned $50,000. Solo designers, who freelance or work independently of a company, reported median earnings of $50,000. Design directors, the creative heads of design firms or in-house corporate design departments, earned $80,000. Graphic designers with business responsibilities for the operation of a firm as owners, partners, or principals earned $90,000.

Related Occupations

Workers in other occupations who design or arrange objects, materials, or interiors to enhance their appearance and function include artists and related workers, architects, except landscape and naval, engineers, landscape architects, and photographers. Some computer-related occupations require design skills, including computer software engineers and desktop publishers.

Sources of Additional Information

For general information about art and design and a list of accredited college-level programs, contact:

- National Association of Schools of Art and Design, 11250 Roger Bacon Dr., Suite 21, Reston, VA 20190. Internet: <http://www.arts-accredit.org> (Search term: nasad)
- For information on industrial design careers and a list of academic programs in industrial design, write to: Industrial Designers Society of America, 1142 Walker Rd., Great Falls, VA 22066. Internet: <http://www.idsa.org>
- For information about graphic design careers, contact: American Institute of Graphic Arts, 164 Fifth Ave., New York, NY 10010. Internet: <http://www.aiga.org>

DESKTOP PUBLISHERS

Significant Points

- Desktop publishers rank among the 10 fastest growing occupations.
- Most jobs are in firms that handle commercial or business printing, and in newspaper plants.
- Although formal training is not always required, those with certification or degrees will have the best job opportunities.

Nature of the Work

Using computer software, desktop publishers format and combine text, numerical data, photographs, charts, and other visual graphic elements to produce publication-ready material. Depending on the nature of a particular project, desktop publishers may write and edit text, create graphics to accompany text, convert photographs and drawings into digital images and then manipulate those images, design page layouts, create proposals, develop presentations and advertising campaigns, typeset and do color separation, and translate electronic information onto film or other traditional forms. Materials produced by desktop publishers include books, business cards, calendars, magazines, newsletters and newspapers, packaging, slides, and tickets. As companies have brought the production of marketing, promotional, and other kinds of materials in-house, they increasingly have employed people who can produce such materials.

Desktop publishers use a keyboard to enter and select formatting specifics such as size and style of type, column width, and spacing, and store them in the computer. The computer then displays and arranges columns of type on a video display terminal or computer monitor. An entire newspaper, catalog, or book page, complete with artwork and graphics, can be created on the screen exactly as it will appear in print. Operators transmit the pages for production either into film and then into printing plates, or directly into plates.

Desktop publishing is a rapidly changing field that encompasses a number of different kinds of jobs. Personal computers enable desktop publishers to perform publishing tasks that would otherwise require complicated equipment and human effort. Advances in computer software and printing technology continue to change and enhance desktop publishing work. Instead of receiving simple typed text from customers, desktop publishers get the material on a computer disk. Other innovations in desktop publishing work include digital color page makeup systems, electronic page layout systems, and off-press color proofing systems. Moreover, because most materials today often are published on the Internet, desktop publishers may need to know electronic publishing technologies, such as Hypertext Markup Language (HTML), and may be responsible for converting text and graphics to an Internet-ready format.

Typesetting and page layout have been affected by the technological changes shaping desktop publishing. Increasingly, desktop publishers use computers to do much of the typesetting and page layout work formerly done by prepress workers, posing new challenges for the printing industry. The old "hot type" method of text composition—which used molten lead to create individual letters, paragraphs, and full pages of text—is nearly extinct. Today, composition work is primarily done with computers. Improvements in desktop publishing software also allow customers to do much more of their own typesetting.

Desktop publishers use scanners to capture photographs, images or art as digital data that can be incorporated directly into electronic page layouts or further manipulated using computer software. The desktop publisher then can correct for mistakes or compensate for deficiencies in the original color print or transparency.

Digital files are used to produce printing plates. Like photographers and multimedia artists and animators, desktop publishers also can create special effects or other visual images using film, video, computers, or other electronic media.

Depending on the establishment employing these workers, desktop publishers also may be referred to as publications specialists, electronic publishers, DTP operators, desktop publishing editors, electronic prepress technicians, electronic publishing specialists, image designers, typographers, compositors, layout artists, and web publications designers.

Working Conditions

Desktop publishers usually work in clean, air-conditioned office areas with little noise. Desktop publishers usually work an 8-hour day, 5 days a week. Some workers—particularly those self-employed—work night shifts, weekends, and holidays.

Desktop publishers often are subject to stress and the pressures of short deadlines and tight work schedules.

Like other workers who spend long hours working in front of a computer monitor, they may be susceptible to eyestrain, back discomfort, and hand and wrist problems.

Employment

Desktop publishers held about 38,000 jobs in 2000. Nearly all worked in the printing and publishing industries. About 1,000 desktop publishers were self-employed.

Most desktop publishing jobs were found in firms that handle commercial or corporate printing, and in newspaper plants. Commercial printing firms print a wide range of products—newspaper inserts, catalogs, pamphlets, and advertisements—while business form establishments print material such as sales receipts. A large number of desktop publishers also were found in printing trade services firms. Establishments in printing trade services typically perform custom compositing, plate making, and related prepress services. Others work printing or publishing materials "in-house" or "in-plant" for business services firms, government agencies, hospitals, or universities, typically in a reproduction or publications department that operates within the organization.

The printing and publishing industry is one of the most geographically dispersed in the United States, and desktop publishing jobs are found throughout the country. However, job prospects may be best in large metropolitan cities.

Training, Other Qualifications, and Advancement

Most workers qualify for jobs as desktop publishers by taking classes or completing certificate programs at vocational schools, universities and colleges, or via the Internet. Programs range in length, but the average non-degree certification training program takes approximately 1 year. However, some desktop publishers train on the job to develop the necessary skills. The length of training on the job varies by company. An internship or part-time desktop publishing assignment is another way to gain experience as a desktop publisher.

Students interested in pursuing a career in desktop publishing also may obtain an associate degree in applied science or a bachelor's degree in graphic arts, graphic communications or graphic design. Graphic arts programs are a good way to learn about desktop publishing software used to format pages, assign type characteristics, and import text and graphics into electronic page layouts to produce printed materials such as advertisements, brochures, newsletters, and forms. Applying this knowledge of graphic arts techniques and computerized typesetting usually is intended for students who may eventually move into management positions, while 2-year associate degree programs are designed to train skilled workers. Students also develop finely tuned skills in typography, print mediums, packaging, branding and identity, Web design and motion graphics. These programs teach print and graphic design fundamentals and provide an extensive background in imaging, prepress, print reproduction, and emerging media. Courses in other aspects of printing also are available at vocational-technical institutes, industry-sponsored update and retraining programs, and private trade and technical schools.

Although formal training is not always required, those with certification or degrees will have the best job opportunities. Most employers prefer to hire people who have at least a high school diploma, possess good communication skills, basic computer skills, and a strong work ethic. Desktop publishers should be able to deal courteously with people because in small shops they may have to take customer orders. They also may add, subtract, multiply, divide, and compute ratios to estimate job costs. Persons interested in working for firms using advanced printing technology need to know the basics of electronics and computers.

Desktop publishers need good manual dexterity, and they must be able to pay attention to detail and work independently. Good eyesight, including visual acuity, depth perception, field of view, color vision, and the ability to focus quickly, also are assets. Artistic ability often is a plus. Employers also seek persons who are even-tempered and adaptable—important qualities for workers who often must meet deadlines and learn how to operate new equipment.

Workers with limited training and experience may start as helpers. They begin with instruction from an experienced desktop publisher and advance based on their demonstrated mastery of skills at each level. All workers should expect to be retrained from time to time to handle new, improved software and equipment.

As workers gain experience, they advance to positions with greater responsibility. Some move into supervisory or management positions. Other desktop publishers may start their own company or work as an independent consultant, while those with more artistic talent and further education may find opportunities in graphic design or commercial art.

Job Outlook

Employment of desktop publishers is expected to grow much faster than the average for all occupations through 2010, as more page layout and design work is performed in-house using computers and sophisticated publishing software. Desktop publishing is replacing much of the prepress work done by compositors and typesetters, enabling organizations to reduce costs while increasing production speeds. Many new jobs for desktop publishers are expected to emerge in commercial printing and publishing establishments. However, more companies also are turning to in-house desktop publishers, as computers with elaborate text and graphics capabilities have become common, and desktop publishing software has become cheaper and easier to use. In addition to employment growth, many job openings for desktop publishers also will result from the need to replace workers who move into managerial positions, transfer to other occupations, or who leave the labor force.

Printing and publishing costs represent a significant portion of a corporation's expenses, no matter the industry, and corporations are finding it more profitable to print their own newsletters and other reports than to send them out to trade shops. Desktop publishing reduces the time needed to complete a printing job, and allows commercial printers to make inroads into new markets that require fast turnaround.

Most employers prefer to hire experienced desktop publishers. As more people gain desktop publishing experience, however, competition for jobs may increase. Among persons without experience, opportunities should be best for those with computer backgrounds who are certified or who have completed postsecondary programs in desktop publishing or graphic design. Many employers prefer graduates of these programs because the comprehensive training they receive helps them learn the page layout process and adapt more rapidly to new software and techniques.

Earnings

Earnings for desktop publishers vary according to level of experience, training, location, and size of firm.

Median annual earnings of desktop publishers were $30,600 in 2000. The middle 50 percent earned between $22,890 and $40,210. The lowest 10 percent earned less than $17,800, and the highest 10 percent earned more than $50,920 a year. Median annual earnings in the industries employing the largest numbers of these workers in 2000 are shown below:

Commercial printing	$30,940
Newspapers	$24,520

Related Occupations

Desktop publishers use artistic and editorial skills in their work. These skills also are essential for artists and related workers; designers; news analysts, reporters, and correspondents; public relations specialists; writers and editors; and prepress technicians and workers.

Sources of Additional Information

Details about apprenticeship and other training programs may be obtained from local employers such as newspapers and printing shops, or from local offices of the State employment service.

For information on careers and training in printing, desktop publishing, and graphic arts, write to:

- Graphic Communications Council, 1899 Preston White Dr., Reston, VA 20191. Internet: <http://www.npes.org>
- Graphic Arts Technical Foundation, 200 Deer Run Rd., Sewickley, PA 15143. Internet: <http://www.gatf.org>

For information on benefits and compensation in desktop publishing, write to:

- Printing Industries of America, Inc., 100 Daingerfield Rd., Alexandria, VA 22314. Internet: <http://www.gain.org>

PHOTOGRAPHERS

Significant Points

- Technical expertise, a "good eye," imagination, and creativity are essential.
- Only the most skilled and talented who have good business sense maintain long-term careers.
- More than half of all photographers are self-employed, a much higher proportion than the average for all occupations.

Nature of the Work

Photographers produce and preserve images that paint a picture, tell a story, or record an event. To create commercial quality photographs, photographers need both technical expertise and creativity. Producing a successful picture requires choosing and presenting a subject to achieve a particular effect and selecting the appropriate equipment. For example, photographers may enhance the subject's appearance with lighting or draw attention to a particular aspect of the subject by blurring the background.

Today, many cameras adjust settings like shutter speed and aperture automatically. They also let the photographer adjust these settings manually, allowing greater creative and technical control over the picture-taking process. In addition to automatic and manual cameras, photographers use an array of film, lenses, and equipment—from filters, tripods, and flash attachments to specially constructed lighting equipment.

Photographers use either a traditional camera or a newer digital camera that electronically records images.

Some photographers specialize in areas such as portrait, commercial and industrial, scientific, news, or fine arts photography. Industrial photographers often take pictures of equipment, machinery, products, workers, and company officials. The pictures then are used for analyzing engineering projects, publicity, or as records of equipment development or deployment, such as placement of an offshore rig. This photography frequently is done on location.

Scientific photographers photograph a variety of subjects to illustrate or record scientific or medical data or phenomena, using knowledge of scientific procedures. They typically possess additional knowledge in areas such as engineering, medicine, biology, or chemistry.

News photographers, also called photojournalists, photograph newsworthy people; places; and sporting, political, and community events for newspapers, journals, magazines, or television. Some news photographers are salaried staff; others are self-employed and are known as freelance photographers.

Self-employed, or freelance, photographers may license the use of their photographs through stock photo agencies or contract with clients or agencies to provide photographs as necessary. Stock agencies grant magazines and other customers the right to purchase the use of photographs, and, in turn, pay the photographer on a commission basis. Stock photo agencies require an application from the photographer and a sizable portfolio. Once accepted, a large number of new submissions usually are required from the photographer each year.

Working Conditions

Working conditions for photographers vary considerably. Photographers employed in government and advertising agencies usually work a 5-day, 40-hour week. On the other hand, news photographers often work long, irregular hours and must be available to work on short notice. Many photographers work part time or variable schedules.

News and commercial photographers frequently travel locally, stay overnight on assignments, or travel to distant places for long periods.

Some photographers work in uncomfortable, or even dangerous surroundings, especially news photographers covering accidents, natural disasters, civil unrest, or military conflicts. Many photographers must wait long hours in all kinds of weather for an event to take place and stand or walk for long periods while carrying heavy equipment. News photographers often work under strict deadlines.

Self-employment allows for greater autonomy, freedom of expression, and flexible scheduling. However, income can be uncertain and the continuous, time-consuming search for new clients can be stressful. Some self-employed photographers hire assistants who help seek out new business.

Employment

Photographers held about 131,000 jobs in 2000. More than half were self-employed, a much higher proportion than the average for all occupations. Some self-employed photographers contracted with advertising agencies, magazines, or others to do individual projects at a predetermined fee, while others operated portrait studios or provided photographs to stock photo agencies.

Most salaried photographers worked in portrait or commercial photography studios. Newspapers, magazines, television broadcasters, advertising agencies, and government agencies employed most of the others. Most photographers worked in metropolitan areas.

Training, Other Qualifications, and Advancement

Employers usually seek applicants with a "good eye," imagination, and creativity, as well as a good technical understanding of photography. Entry-level positions in photojournalism, industrial, or scientific photography generally require a college degree in journalism or photography.

Freelance and portrait photographers need technical proficiency, whether gained through a degree program, vocational training, or extensive work experience.

Many universities, community and junior colleges, vocational-technical institutes, and private trade and technical schools offer photography courses. Basic courses in photography cover equipment, processes, and techniques. Bachelor's degree programs, especially those including business courses, provide a well-rounded education. Art schools offer useful training in design and composition.

Individuals interested in photography should subscribe to photographic newsletters and magazines, join camera clubs, and seek summer or part-time employment in camera stores, newspapers, or photo studios.

Photographers may start out as assistants to experienced photographers. Assistants learn to mix chemicals, develop film, print photographs, and the other skills necessary to run a portrait or commercial photography business. Freelance photographers also should develop an individual style of photography in order to differentiate themselves from the competition. Some photographers enter the field by submitting unsolicited photographs to magazines and art directors at advertising agencies. For freelance photographers, a good portfolio of their work is critical.

Photographers need good eyesight, artistic ability, and hand-eye coordination. They should be patient, accurate, and detail-oriented. Photographers should be able to work well with others, as they frequently deal with clients, graphic designers, or advertising and publishing specialists. Increasingly, photographers need to know computer software programs and applications that allow them to prepare and edit images.

Portrait photographers need the ability to help people relax in front of the camera. Commercial and fine arts photographers must be imaginative and original. News photographers not only must be good with a camera, but also must understand the story behind an event so their pictures match the story. They must be decisive in recognizing a potentially good photograph and act quickly to capture it.

Photographers who operate their own businesses, or freelance, need business skills as well as talent. These individuals must know how to prepare a business plan; submit bids; write contracts; hire models, if needed; get permission to shoot on locations that normally are not open to the public; obtain releases to use photographs of people; license and price photographs; secure copyright protection for their work; and keep financial records.

After several years of experience, magazine and news photographers may advance to photography or picture editor positions. Some photographers teach at technical schools, film schools, or universities.

Job Outlook

Photographers can expect keen competition for job openings because the work is attractive to many people. The number of individuals interested in positions as commercial and news photographers usually is much greater than the number of openings. Those who succeed in landing a salaried job or attracting enough work to earn a living by freelancing are likely to be the most creative, able to adapt to rapidly changing technologies, and adept at operating a business. Related work experience, job-related training, or some unique skill or talent—such as a background in computers or electronics—also are beneficial to prospective photographers.

Employment of photographers is expected to increase about as fast as the average for all occupations through 2010. Demand for portrait photographers should increase as the population grows.

Moreover, as the number of electronic versions of magazines, journals, and newspapers grows on the Internet, photographers will be needed to provide digital images.

Employment growth of photographers will be constrained somewhat by the widespread use of digital photography. Besides increasing photographers' productivity, improvements in digital technology will allow individual consumers and businesses to produce, store, and access photographic images on their own. Declines in the newspaper industry will reduce demand for photographers to provide still images for print.

Earnings

Median annual earnings of salaried photographers were $22,300 in 2000. The middle 50 percent earned between $16,790 and $33,020. The lowest 10 percent earned less than $13,760, and the highest 10 percent earned more than $46,890. Median annual earnings in the industries employing the largest numbers of salaried photographers were as follows:

Radio and television broadcasting $29,890
Mailing, reproduction, and stenographic services $29,610
Newspapers $28,660
Photographic studios, portrait $19,290

Salaried photographers—more of whom work full time—tend to earn more than those who are self-employed.

Because most freelance and portrait photographers purchase their own equipment, they incur considerable expense acquiring and maintaining cameras and accessories. Unlike news and commercial photographers, few fine arts photographers are successful enough to support themselves solely through their art.

Related Occupations

Other occupations requiring artistic talent include architects, artists and related workers; designers; and television, video, and motion picture camera operators and editors.

Sources of Additional Information

Career information on photography is available from:

- Professional Photographers of America, Inc., 229 Peachtree St. NE., Suite 2200, Atlanta, GA 30303. Internet: <http://www.ppa.com>
- National Press Photographers Association, Inc., 3200 Croasdaile Dr., Suite 306, Durham, NC 27705. Internet: <http://www.nppa.org>

PUBLIC RELATIONS SPECIALISTS

Significant Points

- Although employment is projected to increase much faster than the average, keen competition is expected for entry-level jobs.
- Opportunities should be best for college graduates who combine a degree in public relations or other communications-related fields with a public relations internship or other related work experience.
- The ability to write and speak well is essential.

Nature of the Work

An organization's reputation, profitability, and even its continued existence can depend on the degree to which its targeted "publics" support its goals and policies. Public relations specialists serve as advocates for businesses, nonprofit associations, universities, hospitals, and other organizations, and build and maintain positive relationships with the public. As managers recognize the growing importance of good public relations to the success of their organizations, they increasingly rely on public relations specialists for advice on the strategy and policy of such programs.

Public relations specialists handle organizational functions such as media, community, consumer, and governmental relations; political campaigns; interest-group representation; conflict mediation; or employee and investor relations. However, public relations is not only "telling the organization's story."

Understanding the attitudes and concerns of consumers, employees, and various other groups also is a vital part of the job. To improve communications, public relations specialists establish and maintain cooperative relationships with representatives of community, consumer, employee, and public interest groups and with representatives from print and broadcast journalism.

Informing the general public, interest groups, and stockholders of an organization's policies, activities, and accomplishments is an important part of a public relations specialist's job. The work also involves keeping management aware of public attitudes and concerns of the many groups and organizations with which they must deal.

Public relations specialists prepare press releases and contact people in the media who might print or broadcast their material. Many radio or television special reports, newspaper stories, and magazine articles start at the desks of public relations specialists. Sometimes the subject is an organization and its policies towards its employees or its role in the community. Often the subject is a public issue, such as health, energy, or the environment.

Public relations specialists also arrange and conduct programs to keep up contact between organization representatives and the public. For example, they set up speaking engagements and often prepare speeches for company officials. These specialists represent employers at community projects; make film, slide, or other visual presentations at meetings and school assemblies; and plan conventions. In addition, they are responsible for preparing annual reports and writing proposals for various projects.

In government, public relations specialists—who may be called press secretaries, information officers, public affairs specialists, or communications specialists—keep the public informed about the activities of government agencies and officials. For example, public affairs specialists in the Department of State keep the public informed of travel advisories and of U.S. positions on foreign issues. A press secretary for a member of Congress keeps constituents aware of the representative's accomplishments.

In large organizations, the key public relations executive, who often is a vice president, may develop overall plans and policies with other executives. In addition, public relations departments employ public relations specialists to write, research, prepare materials, maintain contacts, and respond to inquiries.

People who handle publicity for an individual or who direct public relations for a small organization may deal with all aspects of the job. They contact people, plan and research, and prepare material for distribution. They also may handle advertising or sales promotion work to support marketing.

Working Conditions

Some public relations specialists work a standard 35- to 40-hour week, but unpaid overtime is common. Occasionally, they must be at the job or on call around the clock, especially if there is an emergency or crisis.

Public relations offices are busy places; work schedules can be irregular and frequently interrupted. Schedules often have to be rearranged so that workers can meet deadlines, deliver speeches, attend meetings and community activities, or travel.

Employment

Public relations specialists held about 137,000 jobs in 2000. About 6 out of 10 salaried public relations specialists worked in services industries—management and public relations firms, membership organizations, educational institutions, healthcare organizations, social service agencies, and advertising agencies, for example.

Others worked for communications firms, financial institutions, and government agencies. About 8,600 public relations specialists were self-employed.

Public relations specialists are concentrated in large cities, where press services and other communications facilities are readily available and many businesses and trade associations have their headquarters. Many public relations consulting firms, for example, are in New York, Los Angeles, Chicago, and Washington, DC. There is a trend, however, for public relations jobs to be dispersed throughout the Nation, closer to clients.

Training, Other Qualifications, and Advancement

There are no defined standards for entry into a public relations career. A college degree combined with public relations experience, usually gained through an internship, is considered excellent preparation for public relations work; in fact, internships are becoming vital to obtaining employment. The ability to write and speak well is essential. Many entry-level public relations specialists have a college major in public relations, journalism, advertising, or communications. Some firms seek college graduates who have worked in electronic or print journalism. Other employers seek applicants with demonstrated communications skills and training or experience in a field related to the firm's business—science, engineering, sales, or finance, for example.

Many colleges and universities offer bachelors and postsecondary degrees in public relations, usually in a journalism or communications department. In addition, many other colleges offer at least one course in this field. A common public relations sequence includes courses in public relations principles and techniques; public relations management and administration, including organizational development; writing, emphasizing news releases, proposals, annual reports, scripts, speeches, and related items; visual communications, including desktop publishing and computer graphics; and research, emphasizing social science research and survey design and implementation. Courses in advertising, journalism, business administration, finance, political science, psychology, sociology, and creative writing also are helpful. Specialties are offered in public relations for business, government, and nonprofit organizations.

Many colleges help students gain part-time internships in public relations that provide valuable experience and training. The Armed Forces also can be an excellent place to gain training and experience. Membership in local chapters of the Public Relations Student Society of America (affiliated with the Public Relations Society of America) or the International Association of Business Communicators provides an opportunity for students to exchange views with public relations specialists and to make professional contacts that may help them find a job in the field. A portfolio of published articles, television or radio programs, slide presentations, and other work is an asset in finding a job. Writing for a school publication or television or radio station provides valuable experience and material for one's portfolio.

Creativity, initiative, good judgment, and the ability to express thoughts clearly and simply are essential. Decision-making, problem-solving, and research skills also are important.

People who choose public relations as a career need an outgoing personality, self-confidence, an understanding of human psychology, and an enthusiasm for motivating people. They should be competitive, yet flexible, and able to function as part of a team.

Some organizations, particularly those with large public relations staffs, have formal training programs for new employees. In smaller organizations, new employees work under the guidance of experienced staff members. Beginners often maintain files of material about company activities, scan newspapers and magazines for appropriate articles to clip, and assemble information for speeches and pamphlets. They also may answer calls from the press and public, work on invitation lists and details for press conferences, or escort visitors and clients. After gaining experience, they write news releases, speeches, and articles for publication or design and carry out public relations programs. Public relations specialists in smaller firms usually get all-around experience, whereas those in larger firms tend to be more specialized.

The Public Relations Society of America accredits public relations specialists who have at least 5 years of experience in the field and have passed a comprehensive 6-hour examination (5 hours written, 1 hour oral). The International Association of Business Communicators also has an accreditation program for professionals in the communications field, including public relations specialists. Those who meet all the requirements of the program earn the Accredited Business Communicator designation. Candidates must have at least 5 years of experience in a communication field and pass a written and oral examination. They also must submit a portfolio of work samples demonstrating involvement in a range of communication projects and a thorough understanding of communication planning. Employers may consider professional recognition through accreditation a sign of competence in this field, which could be especially helpful in a competitive job market.

Promotion to supervisory jobs may come as public relations specialists show that they can handle more demanding assignments. In public relations firms, a beginner may be hired as a research assistant or account assistant and be promoted to account executive, account supervisor, vice president, and, eventually, senior vice president. A similar career path is followed in corporate public relations, although the titles may differ. Some experienced public relations specialists start their own consulting firms. . . .

Job Outlook

Keen competition will likely continue for entry-level public relations jobs as the number of qualified applicants is expected to exceed the number of job openings. Many people are attracted to this profession due to the high-profile nature of the work and the relative ease of entry. Opportunities should be best for college graduates who combine a degree in journalism, public relations, advertising, or another communications-related field with a public relations internship or other related work experience. Applicants without the appropriate educational background or work experience will face the toughest obstacles.

Employment of public relations specialists is expected to increase much faster than the average for all occupations through 2010. The need for good public relations in an increasingly competitive business environment should spur demand for public relations specialists in organizations of all sizes. Employment in public relations firms should grow as firms hire contractors to provide public relations services rather than support full-time staff. In addition to employment growth, job opportunities should result from the need to replace public relations specialists who take other jobs or who leave the occupation altogether.

Earnings

Median annual earnings for salaried public relations specialists were $39,580 in 2000. The middle 50 percent earned between $29,610 and $53,620; the lowest 10 percent earned less than $22,780, and the top 10 percent earned more than $70,480. Median annual earnings in the industries employing the largest numbers of public relations specialists in 2000 were:

Management and public relations	$43,690
Local government	$40,760
State government	$39,560
Colleges and universities	$35,080

According to a joint survey conducted by the International Association of Business Communicators and the Public Relations Society of America, the median annual income for a public relations specialist was $39,000 in 1999.

Related Occupations

Public relations specialists create favorable attitudes among various organizations, special interest groups, and the public through effective communication. Other workers with similar jobs include advertising, marketing, promotions, public relations, and sales managers; demonstrators, product promoters, and models; news analysts, reporters, and correspondents; lawyers; and police and detectives involved in community relations.

Sources of Additional Information

A comprehensive directory of schools offering degree programs, a sequence of study in public relations, a brochure on careers in public relations, and a $5 brochure entitled, Where Shall I go to Study Advertising and Public Relations, are available from:

- Public Relations Society of America, Inc., 33 Irving Place, New York, NY 10003-2376. Internet: <http://www.prsa.org>
- For information on accreditation for public relations specialists, contact: International Association of Business Communicators, One Hallidie Plaza, Suite 600, San Francisco, CA 94102. Internet: <http://www.iabc.com>

TEACHERS—MIDDLE AND SECONDARY

Significant Points

- Public school teachers must have at least a bachelor's degree, complete an approved teacher education program, and be licensed.
- Many States offer alternative licensing programs to attract people into teaching, especially for hard-to-fill positions.
- Excellent job opportunities will stem from the large number of teachers expected to retire over the next 10 years, particularly at the secondary school level; job outlook will vary by geographic area and subject specialty.

Nature of the Work

Teachers act as facilitators or coaches, using interactive discussions and "hands-on" learning to help students learn and apply concepts in subjects such as agriculture. As teachers move away from the traditional repetitive drill approaches and rote memorization, they are using more "props" or "manipulatives" to help students understand abstract concepts, solve problems, and develop critical thought processes. As children get older, they use more sophisticated materials such as science apparatus, cameras, or computers.

Many classes are becoming less structured, with students working in groups to discuss and solve problems together. Preparing students for the future workforce is the major stimulus generating the changes in education.

To be prepared, students must be able to interact with others, adapt to new technology, and logically think through problems. Teachers provide the tools and environment for their students to develop these skills.

Middle and secondary school teachers help students delve more deeply into subjects introduced in elementary school and expose them to more information about the world. Middle and secondary school teachers specialize in a specific subject, such as agriculture. They also can teach subjects that are career-oriented. Vocational education teachers instruct and train students to work in a wide variety of fields, such as health care, business, auto repair, communications, and, increasingly, technology. They often teach courses that are in high demand by area employers, who may provide input into the curriculum and offer internships to students.

Teachers may use films, slides, overhead projectors, and the latest technology in teaching, including computers, telecommunication systems, and video discs. Use of computer resources, such as educational software and the Internet, exposes students to a vast range of experiences and promotes interactive learning.

Through the Internet, American students can communicate with students in other countries. Students also use the Internet for individual research projects and information gathering. Computers are used in other classroom activities as well, from helping students solve math problems to learning English as a second language.

Teachers also may use computers to record grades and perform other administrative and clerical duties. They must continually update their skills so that they can instruct and use the latest technology in the classroom.

Teachers often work with students from varied ethnic, racial, and religious backgrounds. With growing minority populations in many parts of the country, it is important for teachers to establish rapport with a diverse student population. Accordingly, some schools offer training to help teachers enhance their awareness and understanding of different cultures.

Teachers design classroom presentations to meet student needs and abilities. They also work with students individually. Teachers plan, evaluate, and assign lessons; prepare, administer, and grade tests; listen to oral presentations; and maintain classroom discipline. They observe and evaluate a student's performance and potential, and increasingly are asked to use new assessment methods. For example, teachers may examine a portfolio of a student's artwork or writing to judge the student's overall progress. They then can provide additional assistance in areas where a student needs help. Teachers also grade papers, prepare report cards, and meet with parents and school staff to discuss a student's academic progress or personal problems.

In addition to classroom activities, teachers oversee study halls and homerooms, supervise extracurricular activities, and accompany students on field trips. They identify physical or mental problems and refer students to the proper resource or agency for diagnosis and treatment. Secondary school teachers occasionally assist students in choosing courses, colleges, and careers. Teachers also participate in education conferences and workshops.

In recent years, site-based management, which allows teachers and parents to participate actively in management decisions, has gained popularity. In many schools, teachers are increasingly involved in making decisions regarding the budget, personnel, textbook choices, curriculum design, and teaching methods.

Working Conditions

Seeing students develop new skills and gain an appreciation of knowledge and learning can be very rewarding.

However, teaching may be frustrating when one is dealing with unmotivated or disrespectful students.

Occasionally, teachers must cope with unruly behavior and violence in the schools. Teachers may experience stress when dealing with large classes, students from disadvantaged or multicultural backgrounds, and heavy workloads. Schools, particularly in inner cities, may be run down and lack the amenities of schools in wealthier communities.

Teachers are sometimes isolated from their colleagues because they work alone in a classroom of students. However, some schools are allowing teachers to work in teams and with mentors to enhance their professional development.

Including school duties performed outside the classroom, many teachers work more than 40 hours a week.

Most teachers work the traditional 10-month school year with a 2-month vacation during the summer. During the vacation break, those on the 10-month schedule may teach in summer sessions, take other jobs, travel, or pursue other personal interests. Many enroll in college courses or workshops to continue their education.

Teachers in districts with a year-round schedule typically work 8 weeks, are on vacation for 1 week, and have a 5-week midwinter break. Preschool teachers working in day care settings often work year round.

Most States have tenure laws that prevent teachers from being fired without just cause and due process. Teachers may obtain tenure after they have satisfactorily completed a probationary period of teaching, normally 3 years. Tenure does not absolutely guarantee a job, but it does provide some security.

Employment

Teachers held about 3.8 million jobs in 2000. Of those, about 1.5 million were elementary school teachers, 1.1 million were secondary school,

590,000 were middle school, 423,000 were preschool, and 175,000 were kindergarten teachers. Approximately 15 percent of elementary, middle, and secondary school teachers work for private schools. Preschool facilities are often located in schools, religious institutions, and workplaces in which employers provide day care for their employees' children. Employment of teachers is distributed geographically, much the same as the population.

Training, Other Qualifications, and Advancement

All 50 States and the District of Columbia require public school teachers to be licensed. Licensure is not required for teachers in private schools. Usually, the State board of education or a licensure advisory committee grants licensure.

Requirements for regular licenses to teach kindergarten through grade 12 vary by State. However, all States require general education teachers to have a bachelor's degree and to have completed an approved teacher-training program with a prescribed number of subject and education credits as well as supervised practice teaching. About one-third of the States also require technology training as part of the teacher certification process. A number of States require specific minimum grade point averages for teacher licensure. Other States require teachers to obtain a master's degree in education, which involves at least 1 year of additional coursework beyond the bachelor's degree, with a specialization in a particular subject.

Almost all States require applicants for teacher licensure to be tested for competency in basic skills such as reading, writing, teaching, and subject matter proficiency. Most States require continuing education for renewal of the teacher's license. Many States have reciprocity agreements that make it easier for teachers licensed in one State to become licensed in another.

Increasingly, States are moving towards implementing performance-based standards for licensure, which require passing a rigorous comprehensive teaching examination to obtain a provisional license. Teachers must then demonstrate satisfactory teaching performance over an extended period to obtain a full license.

Many States offer alternative teacher licensure programs for people who have bachelor's degrees in the subject they will teach, but lack the necessary education courses required for a regular license. Alternative licensure programs originally were designed to ease teacher shortages in certain subjects, such as mathematics and science. The programs have expanded to attract other people into teaching, including recent college graduates and mid-career changers. In some programs, individuals begin teaching quickly under provisional licensure. After working under the close supervision of experienced educators for 1 or 2 years while taking education courses outside school hours, they receive regular licensure if they have progressed satisfactorily.

Under other programs, college graduates who do not meet licensure requirements take only those courses that they lack, and then become licensed. This may take 1 or 2 semesters of full-time study. States may issue emergency licenses to individuals who do not meet requirements for a regular license when schools cannot attract enough qualified teachers to fill positions. Teachers who need licensure may enter programs that grant a master's degree in education, as well as a license.

In many States, vocational teachers have many of the same requirements for teaching as their academic counterparts. However, since knowledge and experience in a particular field are the most important criteria for the job, some States will license vocational education teachers without a bachelor's degree, provided they can demonstrate expertise in their field.

Teacher education programs are now required to include classes in the use of computers and other technologies to maintain accreditation. Most programs require students to perform a student teaching internship.

Many States now offer professional development schools, which are partnerships between universities and elementary or secondary schools. Students enter these 1-year programs after completion of their bachelor's degree. Professional development schools merge theory with practice and allow the student to experience a year of teaching first-hand, with professional guidance.

In addition to being knowledgeable in their subject, teachers must have the ability to communicate, inspire trust and confidence, and motivate students, as well as understand their educational and emotional needs.

Teachers must be able to recognize and respond to individual differences in students, and employ different teaching methods that will result in higher student achievement. They should be organized, dependable, patient, and creative. Teachers also must be able to work cooperatively and communicate effectively with other teaching staff, support staff, parents, and other members of the community.

With additional preparation, teachers may move into positions as school librarians, reading specialists, curriculum specialists, or guidance counselors. Teachers in kindergarten through grade 12 may become administrators or supervisors, although the number of these positions is limited and competition can be intense.

In some systems, highly qualified, experienced teachers can become senior or mentor teachers, with higher pay and additional responsibilities. They guide and assist less experienced teachers while keeping most of their own teaching responsibilities. Preschool teachers usually work their way up from assistant teacher, to teacher, then to lead teacher—who may be responsible for instruction of several classes—and finally to director of the center.

Job Outlook

Job opportunities for teachers over the next 10 years should be excellent, attributable mostly to the large number of teachers expected to retire. Although employment of preschool, kindergarten, elementary, middle, and secondary school teachers is expected to increase about as fast as the average for all occupations, a large proportion will be eligible to retire by 2010, creating many vacancies, particularly at the secondary school level.

Intense competition for good teachers is already under way among employers in many locations, with schools luring teachers from other States and districts with bonuses and higher pay.

Overall enrollments through 2010, a key factor in the demand for teachers, are projected to rise slowly, resulting in average employment growth for all teachers from preschool to secondary grades. However, projected enrollments vary by region. States in the South and West—particularly California, Texas, Arizona, and Georgia—will experience large enrollment increases, while States in the Northeast and Midwest may experience declines. Projected enrollments also differ by grade, with enrollments rising moderately in grades 9 through 12, while remaining fairly steady for all other grades over the 2000–10 period.

The job market for teachers also continues to vary by school location and by subject specialty. Many inner cities—often characterized by overcrowded, ill-equipped schools and higher than average poverty rates—and rural areas—characterized by their remote location and relatively low salaries—have difficulty attracting enough teachers, so job prospects should be better in these areas than in suburban districts. Teachers who are geographically mobile and who obtain licensure in more than one subject should have a distinct advantage in finding a job. Increasing enrollments of minorities, coupled with a shortage of minority teachers, should cause efforts to recruit minority teachers to intensify.

The number of teachers employed also is dependent on State and local expenditures for education and enactment of legislation to increase the quality of education. Because of a shortage of teachers in certain locations and in anticipation of the loss of a number of teachers to retirement, many States are implementing policies that will encourage more students to become teachers. Some are giving large signing bonuses that are distributed over the teacher's first few years of teaching. Some are expanding State scholarships; issuing loans for moving expenses; and implementing loan-forgiveness programs, allowing education majors with at least a B average to receive State-paid tuition so long as they agree to teach in the State for 4 years.

The supply of teachers also is expected to increase in response to reports of improved job prospects, more teacher involvement in school policy, and greater public interest in education. In recent years, the total number of bachelor's and master's degrees granted in education has steadily increased. In addition, more teachers will be drawn from a reserve pool of career changers, substitute teachers, and teachers completing alternative certification programs, relocating to different schools, and re-entering the workforce.

Earnings

Median annual earnings of kindergarten, elementary, middle, and secondary school teachers ranged from $37,610 to $42,080 in 2000; the lowest 10 percent earned $23,320 to $28,460; the top 10 percent earned $57,590 to $64,920. Median earnings for preschool teachers were $17,810.

According to the American Federation of Teachers, beginning teachers with a bachelor's degree earned an average of $27,989 in the 1999–2000 school year. The estimated average salary of all public elementary and secondary school teachers in the 1999–2000 school year was $41,820. Private school teachers generally earn less than public school teachers do.

In 1999, more than half of all public school teachers belonged to unions—mainly the American Federation of Teachers and the National Education Association—that bargain with school systems over wages, hours, and the terms and conditions of employment.

Teachers can boost their salary in a number of ways. In some schools, teachers receive extra pay for working with students in extracurricular activities. Getting a master's degree or national certification often results in a raise in pay, as does acting as a mentor teacher. Some teachers earn extra income during the summer teaching summer school or performing other jobs in the school system.

Related Occupations

Preschool, kindergarten, elementary, middle, and secondary school teaching requires a variety of skills and aptitudes, including a talent for working with children; organizational, administrative, and record keeping abilities; research and communication skills; the power to influence, motivate, and train others; patience; and creativity. Workers in other occupations requiring some of these aptitudes include teachers-postsecondary; counselors;

teacher assistants; education administrators; librarians; childcare workers; public relations specialists; social workers; and athletes, coaches, umpires, and related workers.

Sources of Additional Information

Information on licensure or certification requirements and approved teacher training institutions is available from local school systems and State departments of education.

Information on the teaching profession and on how to become a teacher can be obtained from:

- Recruiting New Teachers, Inc., 385 Concord Ave., Suite 103, Belmont, MA 02478. Internet: <http://www.rnt.org>. This organization also sponsors another Internet site that provides helpful information on becoming a teacher: <http://www.recruitingteachers.org>

Information on teachers' unions and education-related issues may be obtained from:

- American Federation of Teachers, 555 New Jersey Ave. NW., Washington, DC 20001. Internet: <http://www.aft.org>
- National Education Association, 1201 16th St. NW., Washington, DC 20036. Internet: <http://www.nea.org>

A list of institutions with accredited teacher education programs can be obtained from:

- National Council for Accreditation of Teacher Education, 2010 Massachusetts Ave. NW., Suite 500, Washington, DC 20036. Internet: <http://www.ncate.org>

For information on careers in educating children and issues affecting preschool teachers, contact:

- National Association for the Education of Young Children, 1509 16th St. NW., Washington, DC 20036. Internet: <http://www.naeyc.org>
- Association for Childhood Education International, 17904 Georgia Ave., Suite 215, Olney, MD 20832-2277. Internet: <http://www.acei.org>

For eligibility requirements and a description of the Child Development Associate credential, contact:

- Council for Early Childhood Professional Recognition, 2460 16th St. NW., Washington, DC 20009. Internet: <http://www.cdacouncil.org>

TELEVISION, VIDEO, AND MOTION PICTURE CAMERA OPERATORS AND EDITORS

Significant Points

- Technical expertise, a "good eye," imagination, and creativity are essential.
- Keen competition for job openings is expected, because many talented peopled are attracted to the field.
- About one-fourth of camera operators are self-employed.

Nature of the Work

Television, video, and motion picture camera operators produce images that tell a story, inform or entertain an audience, or record an event. Film and video editors edit soundtracks, film, and video for the motion picture, cable, and broadcast television industries. Some camera operators do their own editing.

Making commercial quality movies and video programs requires technical expertise and creativity. Producing successful images requires choosing and presenting interesting material, selecting appropriate equipment, and applying a good eye and steady hand to assure smooth natural movement of the camera.

Camera operators use television, video, or motion picture cameras to shoot a wide range of subjects, including television series, studio programs, news and sporting events, music videos, motion pictures, documentaries, and training sessions. Some film or videotape private ceremonies and special events. Those who record images on videotape are often called videographers. Many are employed by independent television stations, local affiliates, large cable and television networks, or smaller, independent production companies. Studio camera operators work in a broadcast studio and usually videotape their subjects from a fixed position.

Working Conditions

Working conditions for camera operators and editors vary considerably. Those employed in government, television and cable networks, and advertising agencies usually work a 5-day, 40-hour week.

Some camera operators work in uncomfortable, or even dangerous surroundings, covering accidents, natural disasters, civil unrest, or military conflicts. Many camera operators must wait long hours in all kinds of weather for an event to take place and stand or walk for long periods while carrying heavy equipment.

Employment

Television, video, and motion picture camera operators held about 27,000 jobs in 2000; and film and video editors held about 16,000. One-fourth of camera operators were self-employed. Some self-employed camera operators contracted with television networks, documentary or independent filmmakers, advertising agencies, or trade show or convention sponsors to do individual projects for a predetermined fee, often at a daily rate.

Most salaried camera operators were employed by television broadcasting stations or motion picture studios. Most camera operators and editors worked in metropolitan areas.

Training, Other Qualifications, and Advancement

Employers usually seek applicants with a "good eye," imagination, and creativity, as well as a good technical understanding of camera operation. Camera operators and editors usually acquire their skills through on-the-job training or formal postsecondary training at vocational schools, colleges, universities, or photographic institutes.

Formal education may be required for some positions.

Many universities, community and junior colleges, vocational-technical institutes, and private trade and technical schools offer courses in camera operation and videography. Basic courses cover equipment, processes, and techniques. Bachelor's degree programs, especially those including business courses, provide a well-rounded education.

Individuals interested in camera operations should subscribe to videographic newsletters and magazines, join clubs, and seek summer or part-time employment in cable and television networks, motion picture studios, or camera and video stores.

Camera operators in entry-level jobs learn to set up lights, cameras, and other equipment. They may receive routine assignments requiring camera adjustments or decisions on what subject matter to capture. Camera operators in the film and television industries usually are hired for a project based on recommendations from individuals such as producers, directors of photography, and camera assistants from previous projects, or through interviews with the producer.

Camera operators need good eyesight, artistic ability, and hand-eye coordination. They should be patient, accurate, and detail-oriented. Camera operators also should have good communication skills, and, if needed, the ability to hold a camera by hand for extended periods.

Camera operators who operate their own businesses, or freelance, need business skills as well as talent. These individuals must know how to submit bids; write contracts; get permission to shoot on locations that normally are not open to the public; obtain releases to use film or tape of people; price their services; secure copyright protection for their work; and keep financial records.

With increased experience, operators may advance to more demanding assignments or positions with larger or network television stations.

Job Outlook

Camera operators and editors can expect keen competition for job openings because the work is attractive to many people. The number of individuals interested in positions as videographers and movie camera operators usually is much greater than the number of openings. Those who succeed in landing a salaried job or attracting enough work to earn a living by freelancing are likely to be the most creative, highly motivated, able to adapt to rapidly changing technologies, and adept at operating a business. Related work experience or job-related training also are beneficial to prospective camera operators.

Employment of camera operators and editors is expected to grow faster than the average for all occupations through 2010. Rapid expansion of the entertainment market, especially motion picture production and distribution, will spur growth of camera operators. In addition, computer and Internet services provide new outlets for interactive productions. Camera operators will be needed to film made-for-the-Internet broadcasts such as live music videos, digital movies, sports, and general information or entertainment programming. These images can be delivered directly into the home either on compact discs or over the Internet. Modest growth also is expected in radio and television broadcasting.

Earnings

Median annual earnings for television, video, and motion picture camera operators were $27,870 in 2000. The middle 50 percent earned between $19,230 and $44,150. The lowest 10 percent earned less than $14,130, and the highest 10 percent earned more than $63,690. Median annual earnings were $31,560 in motion picture production and services and $23,470 in radio and television broadcasting.

Median annual earnings for film and video editors were $34,160 in 2000. The middle 50 percent earned between $24,800 and $52,000. The lowest 10 percent earned less than $18,970, and the highest 10 percent earned more than $71,280. Median annual earnings were $36,770 in motion picture production and services, the industry employing the largest numbers of film and video editors.

Many camera operators who work in film or video are freelancers; their earnings tend to fluctuate each year. Because most freelance camera operators purchase their own equipment, they incur considerable expense acquiring and maintaining cameras and accessories.

Related Occupations

Related arts and media occupations include artists and related workers, broadcast and sound engineering technicians and radio operators, designers, and photographers.

Sources of Additional Information

Information about career and employment opportunities for camera operators and film and video editors is available from local offices of State employment service agencies, local offices of the relevant trade unions, and local television and film production companies who employ these workers.

WRITERS AND EDITORS

Significant Points

- Most jobs require a college degree either in the liberal arts—communications, journalism, and English are preferred—or a technical subject for technical writing positions.
- Competition is expected to be less for lower paying, entry-level jobs at small daily and weekly newspapers, trade publications, and radio and television broadcasting stations in small markets.
- Persons who fail to gain better paying jobs or earn enough as independent writers usually are able to transfer readily to communications-related jobs in other occupations.

Nature of the Work

Writers and editors communicate through the written word. Writers and editors generally fall into one of three categories. Writers and authors develop original fiction and nonfiction for books, magazines and trade journals, newspapers, online publications, company newsletters, radio and television broadcasts, motion pictures, and advertisements. Technical writers develop scientific or technical materials, such as scientific and medical reports, equipment manuals, appendices, or operating and maintenance instructions. They also may assist in layout work.

Editors select and prepare material for publication or broadcast and review and prepare a writer's work for publication or dissemination.

Nonfiction writers either select a topic or are assigned one, often by an editor or publisher. Then, they gather information through personal observation, library and Internet research, and interviews. Writers select the material they want to use, organize it, and use the written word to express ideas and convey information. Writers also revise or rewrite sections, searching for the best organization or the right phrasing.

Copy writers prepare advertising copy for use by publication or broadcast media, or to promote the sale of goods and services. Newsletter writers produce information for distribution to association members, corporate employees, organizational clients, or the public. Writers and authors also prepare speeches.

Technical writers put scientific and technical information into easily understandable language. They prepare scientific and technical reports, operating and maintenance manuals, catalogs, parts lists, assembly instructions, sales promotion materials, and project proposals. They also plan and edit technical reports and oversee preparation of illustrations, photographs, diagrams, and charts. Science and medical writers prepare a range of formal documents presenting detailed information on the physical or medical sciences. They impart research findings for scientific or medical professions, organize information for advertising or public relations needs, and interpret data and other information for a general readership.

Many writers prepare material directly for the Internet. For example, they may write for electronic newspapers or magazines, create short fiction, or produce technical documentation only available online. In addition, they may write the text of Web sites. These writers should be knowledgeable about graphic design, page layout and desktop publishing software. Additionally, they should be familiar with interactive technologies of the Web so they can blend text, graphics, and sound together.

Freelance writers sell their work to publishers, publication enterprises, manufacturing firms, public relations departments, or advertising agencies. Sometimes, they contract with publishers to write a book or article. Others may be hired on a job-basis to complete specific assignments such as writing about a new product or technique.

Editors review, rewrite, and edit the work of writers. They may also do original writing. An editor's responsibilities vary depending on the employer and type and level of editorial position held. In the publishing industry, an editor's primary duties are to plan the contents of books, technical journals, trade magazines, and other general interest publications. Editors decide what material will appeal to readers, review and edit drafts of books and articles, offer comments to improve the work, and suggest possible titles. Additionally, they oversee the production of the publications.

Major newspapers and newsmagazines usually employ several types of editors. The executive editor oversees assistant editors who have responsibility for particular subjects, such as local news, international news, feature stories, or sports. Executive editors generally have the final say about what stories are published and how they are covered. The managing editor usually is responsible for the daily operation of the news department.

Assignment editors determine which reporters will cover a given story. Copy editors mostly review and edit a reporter's copy for accuracy, content, grammar, and style.

In smaller organizations, like small daily or weekly newspapers or membership newsletter departments, a single editor may do everything or share responsibility with only a few other people. Executive and managing editors typically hire writers, reporters, or other employees. They also plan budgets and negotiate contracts with freelance writers, sometimes called "stringers" in the news industry. In broadcasting companies, program directors have similar responsibilities.

Editors and program directors often have assistants. Many assistants, such as copy editors or production assistants, hold entry-level jobs. They review copy for errors in grammar, punctuation, and spelling, and check copy for readability, style, and agreement with editorial policy. They suggest revisions, such as changing words or rearranging sentences to improve clarity or accuracy. They also do research for writers and verify facts, dates, and statistics. Production assistants arrange page layouts of articles, photographs, and advertising; compose headlines; and prepare copy for printing. Publication assistants who work for publishing houses may read and evaluate manuscripts submitted by freelance writers, proofread printers' galleys, or answer letters about published material. Production assistants on small papers or in radio stations compile articles available from wire services or the Internet, answer phones, and make photocopies.

Most writers and editors use personal computers or word processors. Many use desktop or electronic publishing systems, scanners, and other electronic communications equipment.

Working Conditions

Some writers and editors work in comfortable, private offices; others work in noisy rooms filled with the sound of keyboards and computer printers as well as the voices of other writers tracking down information over the telephone. The search for information sometimes requires travel to diverse workplaces, such as factories, offices, or laboratories, but many have to be content with telephone interviews, the library, and the Internet.

For some writers, the typical workweek runs 35 to 40 hours. However, writers occasionally may work overtime to meet production deadlines. Those who prepare morning or weekend publications and broadcasts work some nights and weekends.

Freelance writers generally work more flexible hours, but their schedules must conform to the needs of the client. Deadlines and erratic work hours, often part of the daily routine for these jobs, may cause stress, fatigue, or burnout.

Changes in technology and electronic communications also affect a writer's work environment. For example, laptops allow writers to work from home or while on the road. Writers and editors who use computers for extended periods may experience back pain, eyestrain, or fatigue.

Employment

Writers and editors held about 305,000 jobs in 2000. About 126,000 jobs were for writers and authors; 57,000 were for technical writers; and 122,000 were for editors. Nearly one-fourth of jobs for writers and editors were salaried positions with newspapers, magazines, and book publishers. Substantial numbers, mostly technical writers, work for computer software firms. Other salaried writers and editors work in educational facilities, advertising agencies, radio and television broadcasting studios, public relations firms, and business and nonprofit organizations, such as professional associations, labor unions, and religious organizations. Some develop publications and technical materials for government agencies or write for motion picture companies.

Jobs with major book publishers, magazines, broadcasting companies, advertising agencies, and public relations firms are concentrated in New York, Chicago, Los Angeles, Boston, Philadelphia, and San Francisco.

Jobs with newspapers, business and professional journals, and technical and trade magazines are more widely dispersed throughout the country.

Thousands of other individuals work as freelance writers, earning some income from their articles, books, and less commonly, television and movie scripts. Most support themselves with income derived from other sources.

Training, Other Qualifications, and Advancement

A college degree generally is required for a position as a writer or editor. Although some employers look for a broad liberal arts background, most prefer to hire people with degrees in communications, journalism, or English.

For those who specialize in a particular area, such as fashion, business, or legal issues, additional background in the chosen field is expected. Knowledge of a second language is helpful for some positions.

Technical writing requires a degree in, or some knowledge about, a specialized field—engineering, business, or one of the sciences, for example. In many cases, people with good writing skills can learn specialized knowledge on the job. Some transfer from jobs as technicians, scientists, or engineers. Others begin as research assistants, or trainees in a technical information department, develop technical communication skills, and then assume writing duties.

Writers and editors must be able to express ideas clearly and logically and should love to write. Creativity, curiosity, a broad range of knowledge, self-motivation, and perseverance also are valuable. Writers and editors must demonstrate good judgment and a strong sense of ethics in deciding what material to publish.

Editors also need tact and the ability to guide and encourage others in their work.

For some jobs, the ability to concentrate amid confusion and to work under pressure is essential. Familiarity with electronic publishing, graphics, and video production equipment increasingly is needed. Online newspapers and magazines require knowledge of computer software used to combine online text with graphics, audio, video, and 3-D animation.

High school and college newspapers, literary magazines, community newspapers, and radio and television stations all provide valuable, but sometimes unpaid, practical writing experience. Many magazines, newspapers, and broadcast stations have internships for students. Interns write short pieces, conduct research and interviews, and learn about the publishing or broadcasting business.

In small firms, beginning writers and editors hired as assistants may actually begin writing or editing material right away. Opportunities for advancement can be limited, however. In larger businesses, jobs usually are more formally structured. Beginners generally do research, fact checking, or copy editing. They take on full-scale writing or editing duties less rapidly than do the employees of small companies. Advancement often is more predictable, though, coming with the assignment of more important articles.

Job Outlook

Employment of writers and editors is expected to increase faster than the average for all occupations through the year 2010. Employment of salaried writers and editors for newspapers, periodicals, book publishers, and nonprofit organizations is expected to increase as demand grows for their publications.

Magazines and other periodicals increasingly are developing market niches, appealing to readers with special interests. In addition, online publications and services are growing in number and sophistication, spurring the demand for writers and editors.

Businesses and organizations are developing newsletters and Internet websites and more companies are experimenting with publishing materials directly for the Internet. Advertising and public relations agencies, which also are growing, should be another source of new jobs. Demand for technical writers and writers with expertise in specialty areas, such as law, medicine, or economics, is expected to increase because of the continuing expansion of scientific and technical information and the need to communicate it to others.

In addition to job openings created by employment growth, many openings will occur as experienced workers retire, transfer to other occupations, or leave the labor force. Replacement needs are relatively high in this occupation; many freelancers leave because they cannot earn enough money.

Despite projections of fast employment growth and numerous replacement needs, the outlook for most writing and editing jobs is expected to be competitive. Many people with writing or journalism training are attracted to the occupation. Opportunities should be best for technical writers and those with training in a specialized field.

Rapid growth and change in the high technology and electronics industries result in a greater need for people to write users' guides, instruction manuals, and training materials. Developments and discoveries in the law, science, and technology generate demand for people to interpret technical information for a more general audience. This work requires people who are not only technically skilled as writers, but also familiar with the subject area. In addition, individuals with the technical skills for working on the Internet may have an advantage finding a job as a writer or editor.

Opportunities for editing positions on small daily and weekly newspapers and in small radio and television stations, where the pay is low, should be better than those in larger media markets.

Some small publications hire freelance copy editors as backup for staff editors or as additional help with special projects. Aspiring writers and editors benefit from academic preparation in another discipline as well, either to qualify them as writers specializing in that discipline or as a career alternative if they are unable to get a job in writing.

Earnings

Median annual earnings for salaried writers and authors were $42,270 in 2000. The middle 50 percent earned between $29,090 and $57,330. The lowest 10 percent earned less than $20,290, and the highest 10 percent earned more than $81,370. Median annual earnings were $26,470 in the newspaper industry.

Median annual earnings for salaried technical writers were $47,790 in 2000. The middle 50 percent earned between $37,280 and $60,000. The lowest 10 percent earned less than $28,890, and the highest 10 percent earned more than $74,360. Median annual earnings in computer and data processing services were $51,220.

Median annual earnings for salaried editors were $39,370 in 2000. The middle 50 percent earned between $28,880 and $54,320. The lowest 10 percent earned less than $22,460, and the highest 10 percent earned more than $73,330. Median annual earnings in the industries employing the largest numbers of editors were as follows:

Computer and data processing services	$45,800
Periodicals	$42,560
Newspapers	$37,560
Books	$37,550

Related Occupations

Writers and editors communicate ideas and information. Other communications occupations include announcers, interpreters and translators; news analysts, reporters, and correspondents; and public relations specialists.

Sources of Additional Information

For information on careers in technical writing, contact:

- Society for Technical Communication, Inc., 901 N. Stuart St., Suite 904, Arlington, VA 22203. Internet: <http://www.stc.org>

For information on union wage rates for newspaper and magazine editors, contact:

- The Newspaper Guild-CWA, Research and Information Department, 501 Third St. NW., Suite 250, Washington, DC 20001.

EDUCATIONAL OPPORTUNITIES IN AGRICULTURE COMMUNICATIONS

Many of the career fields in agricultural communications require a college degree. Colleges throughout the United States offer courses of study in agriculture communications as both a degreed program and a specialization or minor.

Below is a list of universities in the United States and Puerto Rico that offer a bachelor's degree in agricultural communications. This is not a complete list. It does not include colleges that offer a minor or specialization in agricultural communications or technical schools that offer specialized training. If you are interested in an educational institution not listed, contact it directly for information on agricultural and communications programs.

Cal Poly State University

Brock Center for Agricultural Communication
Cal Poly State University
San Luis Obispo, CA 93407
Phone 805-756-2707
Fax 805-756-2799

Cornell University

Cornell University
328 Kennedy Hall
Ithaca, NY 14853
Phone 607-255-9706
Fax 607-254-1322

Kansas State University

Communications
301 Umberger Hall
Kansas State University
Manhattan, KS 66506-3402
Phone 785-532-1163
Fax 785-532-5633

Kansas State University

Department of Communications
301 Umberger Hall
Manhattan, KS 66506
Phone 785-532-3393 (office)
Fax 785-532-5633 FAX

Michigan State University

ANR Education and Communication Systems
409 Agriculture Hall
Michigan State University
East Lansing, MI 48824
Phone 517-355-6580
Fax 517-353-4981

Montana State University

Agriculture Education
Cheever Hall
MSU-Bozeman
Bozeman, MT 59715
Phone 406-994-3201
Fax 406-994-6696

Agriculture and Technology Education
109 Cheever Hall
MSU-Bozeman
Bozeman, MT 56715
Phone 406-994-2132
Fax 406-994-6696

Ohio State University

Department of Human and Community Resource Development
College of Food, Agricultural, and Environmental Sciences
204 Agriculture Administration Building
2120 Fyffe Rd.
The Ohio State University
Columbus, OH 43210-1067
Phone 614-292-4624
Fax 614-292-7007

Oklahoma State University

Agricultural Education, Communications and 4-Youth Development
442 Agricultural Hall
Oklahoma State University
Stillwater, OK 74078-6041
Phone 405-744-0461
FAX 405-744-5176

Purdue University

Agricultural Communication
Room 209, AGAD Building
Purdue University
West Lafayette, IN 47906-1143
Phone 765-494-8406
Fax 765-496-1117

South Dakota State University

Department of Journalism and Mass Communication
Printing/Journalism 219
South Dakota State University
Brookings, SD 57007-0596
Phone 605-688-4171
Fax 605-688-5034

Texas A&M University

Department of Agricultural Education
Texas A&M University
2116 TAMU
College Station, Texas 77843-2116
Phone 979-862-1507
Fax 979-845-6296

Texas Tech University

Agricultural Education and Communications
Box 42131
Lubbock, TX 79409
Phone 806-742-2816
Fax 806-742-2880

University of Arkansas

Agricultural and Extension Education
University of Arkansas
304A Agriculture Building
Fayetteville, Arkansas 72701
Phone 501-575-7435
Fax 501-575-2610

University of Florida

Agricultural Education and Communication
213 Rolfs Hall
University of Florida
Gainesville, FL 32611-0540
Phone 352-392-0502
Fax 352-392-9585

University of Georgia

Agricultural Leadership, Education and Communication
105 Four Towers
University of Georgia
Athens, GA 30602-4355
Phone 706-542-0715
Fax 706-542-0262

University of Illinois

Department of Human and Community Development
University of Illinois
137 Bevier Hall, 905 S. Goodwin Ave.
Urbana, IL 61801
Phone 217-244-6593
Fax 217-244-7877

University of Kentucky

Agricultural Communications
131 Scovell Hall
University of Kentucky
Lexington, KY 40546-0064
Phone 606-257-4657
Fax 606-257-1512

University of Missouri–Columbia

1-98 Agriculture Building
Agricultural Journalism
University of Missouri-Columbia
Columbia, MO 65211
Phone 573-884-7863 (office)
Fax 573-882-8007

University of Nebraska

Agricultural Leadership, Education and Communication
Room 300, Agriculture Hall
University of Nebraska
Lincoln, NE 68583-0709
Phone 402-472-8742
Fax 402-472-3093

University of Puerto Rico

Department of Agricultural Education
University of Puerto Rico
PO Box 9030
Mayaguez, PR 00681-9030
Phone 787-832-4040
Fax 787-834-5265

University of Wisconsin–River Falls

113 Kleinpell Fine Arts
410 South Third Street
River Falls, Wisconsin 54022
Phone 715-425-3175
Fax 715-425-0669

Agricultural Education Department
University of Wisconsin–River Falls
319 Agricultural Science Building
410 South Third Street
River Falls, WI 54022
Phone 715-425-3555
Fax 715-425-3785

Washington State University

P.O. Box 646120
Washington State University
Pullman, WA 99164-6120
Phone 509-335-2899
Fax 509-335-2722

Washington State University

WSU Cooperative Extension
P.O. Box 646242
Pullman, WA 99164-6244
Phone 509-335-2930
Fax 509-335-2863

University of Wyoming

2219 Carey Avenue
Cheyenne, WY 82002
Phone 307-777-7024
Fax 307-777-6593

CONCLUSION

The agricultural communications field is a versatile one, with many job areas: writer, editor, spokesperson, Web designer, photographer, videographer, and a host of others. The main point is that agricultural communicators can work in a number of fields that are related to both communications and agriculture.

Because agriculture is so versatile, it has a need for people who can communicate the latest in scientific discoveries and the promise of new agribusinesses. The demand is great for people who can write, speak, or communicate visually.

GLOSSARY

Agricultural literacy—being knowledgeable about agriculture and the process by which products go from the field to the store.

Alignment—arrangement or position of an object in a straight line or in parallel lines. In layout terms, alignment deals with the placement of objects of text on the right, left, or middle of a page.

Alternative sources—sources that an author or speaker looks to in order to corroborate information found from initial sources.

Aperture—the opening inside a camera lens that allows light to pass through. On more sophisticated lenses, the aperture is adjustable so that the amount of light and the depth of field can be manipulated.

Balance—a harmonious or satisfying arrangement or proportion of parts or elements, as in a design.

Contrast—to illustrate differences when compared. For example, the black words on white paper show a high degree of contrast. Gray words on white paper have less contrast.

Crop—the process of trimming a photograph to make it fit predetermined boundaries.

Digital camera—a camera that records light transmitted through a lens on light-sensitive diodes and then transfers the image to digital storage such as a memory card, floppy disk, or CD. It does not use film.

Editorializing—To present an opinion in a report that is meant to be objective.

E-mail—a system for sending and receiving messages electronically between personal computers.

Exclusive rights—rights granted by freelance photographers and writers that give content users the exclusive right to publish their works.

Extemporaneous speaking—a style of speaking that involves picking a topic at random, having a short amount of time (usually thirty minutes) to prepare, and delivering a short address on the topic.

Fact checking—the process of going over a document to make sure all of the facts presented are accurate.

Feature story—a story that magazines usually run that are typically 1,500 to 2,000 words in length. A feature story generally goes into much more detail than does a news story.

First-time usage—the rights granted by a freelancer to a magazine for usage of artwork of article. First-time usage rights are the same as a lease agreement, where the party using the work pays for one-time use of the work and must pay an additional sum for any subsequent uses.

Freelance—a writer or photographer who sells work to magazines, newspapers, or Web sites without a long-term commitment to any of them.

f-stop—a camera lens aperture setting that corresponds to an f-number. The f-stop setting controls the amount of light and the depth of field of a photograph.

General marketing—The process of marketing a Web site or product to the public at large and not singling out a single group in which to target its marketing message.

Internet—a worldwide network of computers linked by servers that feed the data to users.

Interviewing—the process of talking with someone in order to gain information about a particular topic.

IP address—a unique address given to a computer using the Internet. An IP address is much like a mailing address in that it points traffic to a particular spot on the Internet.

ISO—the standard of measurement in which the sensitivity of film to light is measured. The lower the ISO number, the less sensitive the film is to light, thus requiring longer relative exposure.

Lead—the first paragraph of a news story that tells who, what, where, when, why, and how.

Lens—the optical element on a camera that allows light to pass through and expose the film. The lens focuses the light and limits the amount of light passing through the lens by means of an aperture.

Link—refers to a Web site. A collection of words, pictures, or other objects that, when clicked, take the Web browser to another page on the site or to another Web site altogether.

Masthead—the listing in a magazine or newspaper with information about the publication's staff, operation, and circulation.

News determinants—a set of criteria that writers and editors use to decide if news is fair to write and print.

Plagiarism—a piece of writing that is copied from another source and passed along as if it were the writer's own.

Prepared public speaking—a style of speaking in which the message is prepared ahead of time.

Press release—a document generated by an entity such as a university or corporation that tells of news that affects it.

Proximity—a design term that refers to the position of one element in relation to another.

Query letter—a letter, usually written to a magazine or newspaper editor, that outlines the details of a proposed story idea.

Research—the process of studying something thoroughly so as to present in a detailed, accurate manner.

Rule of thirds—a rule of composition that dictates the placement of a photographic subject in a photograph.

Server—a computer that processes requests for components of Web pages.

Shutter speed—the speed that a camera's shutter moves in order to expose the film to light. Shutter speed is usually expressed in a fraction of a second (e.g., 1/500 or 1/30).

Spokesperson—an advocate who represents someone else's policy or purpose.

Target marketing—a form of marketing that targets a specific segment of the market. For example, a company may focus its targeted marketing on corn growers.

Trends—a pattern that markets and consumers follow that shows an inclination for things they prefer. For example, bell bottom jeans followed a consumer trend.

Unlimited usage—a usage contract that a freelance writer or photographer negotiates in which the client can use the work the freelancer produced as many times as it chooses without further remuneration to the freelancer.

World Wide Web—the complete set of documents residing on all Internet servers that use the HTTP protocol, accessible to users by a point-and-click system.

INDEX